D0061226

Mathematical Games and How to Play Them

Steven Vajda

Dover Publications, Inc.
Mineola, New York

Copyright

Bibliographical Note

This Dover edition, first published in 2008, is an unabridged republication of the edition first published by Ellis Horwood Limited, Chichester, England, in 1992.

Library of Congress Cataloging-in-Publication Data

Vajda, S.
Mathematical games and how to play them / Steven Vajda.
p. cm.
"This Dover edition, first published in 2008, is an unabridged republication of the edition first published by Ellis Horwood Limited, Chichester, England, in 1992."
ISBN-13: 978-0-486-46277-6
ISBN-10: 0-486-46277-3
1. Game theory. I. Title.

QA269.V28 2008
519.3—dc22

2007031299

Manufactured in the United States of America
Dover Publications, Inc., 31 East 2nd Street, Mineola, N.Y. 11501

Table of contents

To

Cedric A. B. Smith

who knows

how recreational mathematics can be.

Preface

People have been playing games for several thousands of years. Chess, Hopscotch, Poker, Noughts and Crosses, Tennis: all these are games. Why do people play them? To assuage their aggressive instincts, or to exercise their muscles? Is it a healthy activity, or just innocent fun?

These are questions to be answered by psychologists, or by sociologists. In this book we are not concerned with such problems. In fact, we are not concerned with just any type of game. We look at games with a relevant mathematical side. Our players are not athletes; they are thinkers. Even concerning mathematical games, we restrict ourselves to a selection.

Our games finish within a finite time — though we cannot always say how long they may last. We exclude games where chance plays a part, such as card games. Moreover, we do not expect our readers to be sophisticated scholars. Many games are too complex to be considered here. Such a game is, for instance, Chess. Many of us might think that this is fortunate. Chess would lose much of its attraction if a perfect theory for winning, or at least for drawing, were known. At present even computers, though they play brilliant games, can be defeated.

Some of our games are based on fairly intricate mathematics, needed to justify the method for solving them, though the method itself can easily be stated. In such cases we have relegated the mathematical argument to an appendix. Appendices deal also with related mathematics, of interest though not directly relevant to our types of games. The mathematics might become relevant in the future, however.

We cover one-person games, and two person games. The former are also called solitaires, but do not include Patience, where chance comes in. Two-person games are also the concern of the theory of games, initiated by John von Neumann. We use it to solve a few games within the context of linear programming, to which an appendix is devoted.

We must add a few words about the names we give to our various games. In some cases such names are traditional, e.g. Nim, or Nine Men's Morris (*A Midsummer Night's Dream*, Act II, Scene ii); in other cases they conceal a pun, e.g. Sylver Coinage. But many got their names because nobody could think of anything better.

We apologize to those who feel that they should have been credited with a discovery in lieu of somebody else. It is well-nigh impossible to reconstruct the history in all cases. In any case, games have been discovered and re-discovered (Wythoff 's game by R. P. Isaacs, or the game of Divisions by D. Gale). Also, games may have different names in different languages, and a number of games have acquired commercially guaranteed names.

We have tried to make the presentation as up-to-date as possible. No doubt new games will have been invented before the reader can see these pages. Recreational mathematics is a healthy and ever-active field. May we all enjoy it.

The Morra-players in the Frontispiece come from Sir Gardner Wilkinson, *The Ancient Egyptians* (1853), 190.* The figures in the text were drawn by N. R. Firth.

*I am grateful to David M. G. Wishart for drawing my attention to this source.

1

Solitaires

Solitaires are games played by a single person. A configuration of pegs, or counters, or coins is given and must be transformed into some other configuration, according to well defined legal moves. Two questions arise:

—Is such a tranformation possible?
—If it is, how is it to be done?

1. GRAPH SOLITAIRE

A configuration is defined by counters on some vertices of a graph. A counter can be moved onto another vertex, if the graph contains an edge between these two vertices. We call a sequence of moves starting from configuration C_1 and producing a configuration C_2 a path between C_1 and C_2. It is represented by a sequence of adjoining edges.

Example 1
Let the graph be a square with vertices A, B, C, and D and let the edges be the four sides of the square and the diagonal AC. A configuration consists of three counters marked 1, 2 and 3 on three of the vertices, with one vertex left uncovered, empty. There are 24 such configurations. A legal move consists of moving a counter to an adjoining vertex, along an edge, provided that adjoining vertex is empty.

We denote the graph by G, and we construct another graph, which Wilson (1974) calls puz (G). This graph has 24 vertices corresponding to the 24 configurations of G, and an edge between vertices V_1 and V_2, if the configuration on G represented by V_1 can be transformed by a single move into the configuration represented by V_2. Puz (G) is drawn in Fig I.1 and we see that there is a path between any two of its vertices. The graph is connected. The 12 vertices corresponding to configurations where one end of the diagonal AC is empty are connected to three other vertices, and the other 12 vertices are connected to two more vertices.

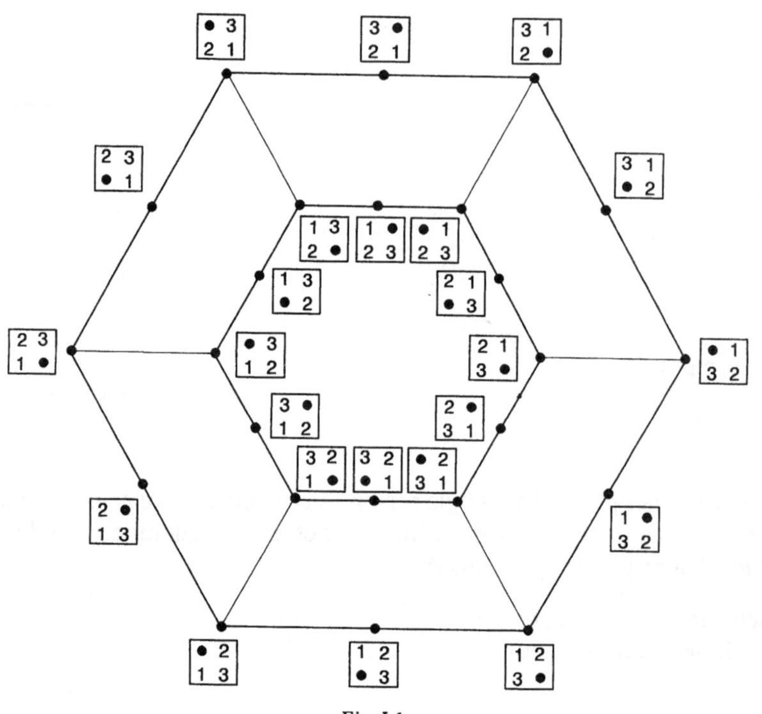

Fig. I.1.

Example II

This example differs from example I by the fact that the diagonal AC of the square is now omitted. Call the resulting square H. If we construct puz (H), we notice that there is now no connecting edge between a vertex of the outer and a vertex of the inner hexagonal component. Considering graphs in general, with one single vertex empty, how many components will the puz-graph have?

Regarding any graph which is finite, non-separable, and simple — that is without loops and multiple edges — Wilson (1974) has shown that, apart from polygons, its puz-graph is connected unless the graph is bipartite. In the latter case, the puz-graph has precisely two components. There is one exception to this rule: the graph with seven vertices, a hexagon and its centre connected by edges to two opposite vertices, has a puz-graph with six components.

As to polygons, it is easy to show that an n-polygon has a puz-graph of $(n-2)!$ components, each with $n(n-1)$ vertices. Curiously, our square in example II is a polygon, and bipartite, so that it is not covered by Wilson's rule. Nevertheless, it has precisely two components.

The grid of 4 by 4 dots with edges between any two vertically or horizontally adjacent dots is also bipartite. It will be the subject of the next section.

2. THE FIFTEEN PUZZLE

This game for one player, also called the Boss Puzzle, and in France Le Jeu du Taquin, was all the craze in the last century, and is still popular. It has also been marketed, under the name of Diablotin, and probably under many other names as well. It is played on a board of four by four squares, equivalent to a grid of sixteen dots in four rows and four columns. Counters, numbered 1 to 15 are placed on fifteen of the squares, one square remaining empty.

We call the configuration in Fig. I.2a the standard configuration, with the numbers, and the empty square, in their standard positions. There are in all 16! (about 2×10^{13}) possible configurations. The reader will not expect us to exhibit them all here.

1	2	3	4
5	6	7	8
9	10	11	12
13	14	15	

(a)

1	2	3	4
5	7	10	8
9	6		11
13	14	12	15

(b)

Fig. I.2.

A move consists of sliding a counter, above or below or to the right or to the left of the empty square, into the latter, thus emptying another square.

It is convenient to number the empty square 16, and to rewrite a configuration as a permutation, reading the numbers from left to right one line after another, as one reads a book. For instance, the configuration in Fig. I.2b would be written

1 2 3 4 5 7 10 8 9 6 16 11 13 14 12 15

The appearance of a number before a smaller one is called an inversion. If the number of inversions is even, then the permutation has even parity; otherwise the parity is odd. The standard configuration, with no inversions, is even. The permutation in Fig. I.2(b) above has 13 inversions: its parity is odd:

(7,6) (10,6) (10,8) (10,9) (8,6) (9,6) (16,11) (16,12)
(16,13) (16,14) (16,15) (13,12) (14,12).

Every move exchanges two elements of the permutation (one of them being 16), so that every move changes parity. For instance, if in the example above we slide counter 10 into the empty square below it, then we obtain an even permutation, with 14 inversions.

Now imagine a configuration with the empty square in its standard position, and a sequence of moves which brings it back into this same position. The other counters might have, or might not have, changed their position; the empty square must have changed its position an even number of times. Hence the new configuration must have the same parity as the old one. Consequently, if we start from any configuration with the empty square in its standard position, and wish to recover the standard configuration, with all numbers in their standard position, then this is only possible if the starting configuration has even parity. Also, starting with the standard configuration, only configurations with even parity can be generated.

Thus, the equality of parity of a starting and a finishing configuration is a necessary condition for the existence of a sequence of moves, a path, between them. It is also a sufficient condition. The following paraphrases the argument in Spitznagel (1967).

It is shown in Appendix II that an even permutation is the product of cycles of degree 3.

To begin with, let the permutation consist of just one such cycle (abc). It is easy to move the counters $a, b, c,$ and the empty square into the lower right-hand corner of the board, with the empty square in its standard position. We then make the following moves:

$$\begin{array}{cc} a & b \\ c & \end{array} \rightarrow \begin{array}{cc} a & \\ c & b \end{array} \rightarrow \begin{array}{cc} & a \\ c & b \end{array} \rightarrow \begin{array}{cc} c & a \\ & b \end{array} \rightarrow \begin{array}{cc} c & a \\ b & \end{array}$$

Now we trace the steps, which have brought us to the lower right hand corner backwards, and $a, b,$ and c will be in their standard positions. The effect of the cycle has been undone.

If a permutation is the product of more than one cycle of degree 3, then we undo the effects of the successive cycles one by one, and all counters will be in their standard positions. (It is easy to see how this proof can be made to fit the case when the empty square is itself one of $a, b,$ or c.)

Of course, we do not suggest that the way described in this existence proof is the best, or even a convenient way of arriving from an even configuration to the standard one. In practice it would be far too cumbersome to keep track of all the steps taken to reach the bottom corner, and to trace them back in the opposite direction. It is better to proceed as follows:

First put 1 into its standard position, then 2, and so on up to 12. This will be found to be easy by moving the empty square around. Having done this, we shall find in the bottom row

(A) 13,14,15,16 or (B) 14,15,13,16 or (C) 15,13,14,16.

No other permutation of 13, 14, 15 can appear, because we assume that we start with an even permutation.

If we have reached (A), then we have finished. In the other two cases we proceed as in Figs I. 3(a) and (b).

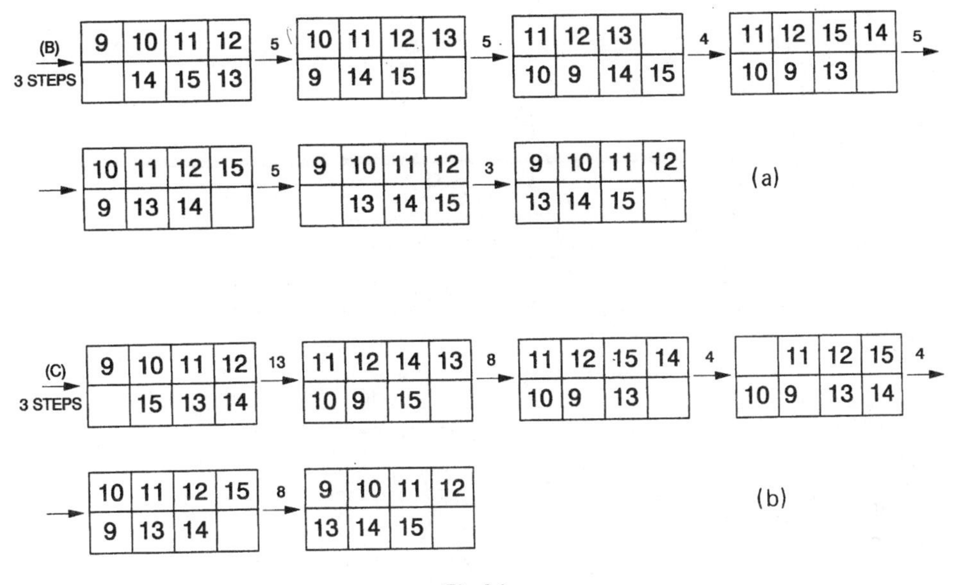

Fig. I.3.

Similar considerations apply to rectangles with r rows and c columns. Liebeck (1971) has considered turning such a rectangle counterclockwise through 90°. He showed by a count of re-arrangements that the counters can be put again into a configuration identical with the original one

—if $c \equiv 3 \pmod 4$ and $r \equiv 0 \pmod 4$ or $r \equiv 3 \pmod 4$
—or if $c \equiv 2 \pmod 4$ if $r \equiv 1 \pmod 4$ or $r \equiv 2 \pmod 4$
—or if $c \equiv 1 \pmod 4$, whatever the value of r is.

If $c \equiv 0 \pmod 4$, the re-arrangement is impossible.

3. TRIANGULAR SOLITAIRE

Consider the triangular board in Fig. I.4 (Hentzel, 1973) which can be extended ad lib. (Ignore the letters, for the time being.) The game is played with counters on the white triangles only. At least one triangle is not covered. Whenever two adjacent triangles (those with a common vertex) contain a counter, while a third adjacent triangle in the same direction is empty, then one counter jumps over another into the empty triangle, and the counter over which the other jumped is removed from the board.

It is required to change a given configuration of counters into another given configuration, of course with fewer counters. The game most often played is that of starting with only one empty triangle, and finishing with all but one triangle empty.

For instance, on the board with triangles 1 to 10 start with triangle 2 the only empty one, and finish with triangle 3 the only one to be covered. This can be done as follows:

7–2, 9–7, 1–4, 7–2, 6–4–1–6, 10–3.

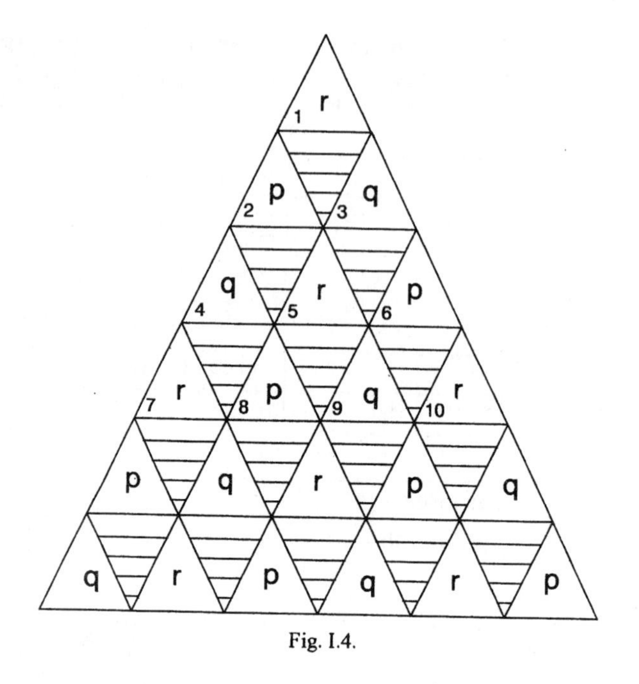

Fig. I.4.

Given the starting and a finishing configuration, a transition by legal moves is not always possible. To explore this, we enter marks *p, q,* or *r* into the triangles (see Fig. I.4). These marks, together with 0 (zero), are subject to rules of combination which we call Nim-addition (in view of their role in the game of Nim (q.v.). Denoting this combination by $\hat{+}$, we have

$$p \hat{+} q = r \quad p \hat{+} r = q \quad q \hat{+} r = p \tag{i}$$

$$p \hat{+} p = 0 \quad q \hat{+} q = 0 \quad r \hat{+} r = 0 \tag{ii}$$

and hence

$$p \hat{+} q \hat{+} r = 0 \tag{iii}$$

We may think of these marks as the numbers 1, 2, 3, 0 in binary notation and their Nim-addition as addition of digits modulo 2 (without carry), thus

$$\begin{array}{ll} p & 01 \\ q & \underline{10} \\ r & 11 \end{array} \qquad \begin{array}{ll} p & 01 \\ r & \underline{11} \\ q & 10 \end{array} \qquad \begin{array}{ll} q & 10 \\ r & \underline{11} \\ p & 01 \end{array}$$

It follows from (i) and (ii) that $p \hat{+} q \hat{+} r = 0$, and the triangles were lettered in such a way that the entries in any three successive triangles add up to 0. If a counter jumps over another counter into an empty triangle, thereby vacating two triangles and covering a third, then the sum of the entries in the former two equals the value in the latter. The Nim-sum of all covered triangles remains the same.

Consequently, any configuration can only change into another with the same Nim-sum of their covered triangles. This is a necessary condition for the possibility of a transformation. However, the condition is not sufficient.

To explore this point of possible transformations further, we compute the Nim-total of all marks in all triangles for triangular boards of n triangles a side. We find that this total is

—r whenever n exceeds a multiple of 3 by 1, and
—0 in all other cases.

If, then, the top triangle marked r an Fig. I.4 is the only empty one, then the Nim-sum of the remaining triangles is 0, whenever n equals 4, or 7, or 10, and so on. Such a configuration cannot possibly be transformed into a configuration with one single triangle not empty, because there is no triangle with the entry 0.

If on the board with four triangles a side the central triangle (with entry r) is empty, no move is possible at all, since there is no triangle from which to jump into it.

The reader might amuse himself by finding possible transformations and ascertaining the smallest number of moves necessary, if consecutive jumps, such as 6–4–1–6 in the example are counted as one move.

4. PEG SOLITAIRE

The English board for this game consists of 33 squares, arranged as in Fig. I.5. The game is played with counters or pegs, on the squares, with at least one square empty. A move consists of a counter jumping horizontally or vertically over an adjacent covered square, into an empty one. The counter jumped over is removed from the board. Thus every move reduces the number of counters in play by one.

		37	47	57		
		36	46	56		
15	25	35	45	55	65	75
14	24	34	44	54	64	74
13	23	33	43	53	63	73
		32	42	52		
		31	41	51		

Fig. I.5.

The general problem can be described as follows: given two configurations, with some squares empty and others covered, how can one of them be transformed into the other? Obviously, this can only be possible if the configuration reached has fewer counters covering squares than the starting configuration. However, even if this is so, a transformation may not be possible. We are therefore, first of all interested in finding a necessary condition for the transformation to be achievable.

For this purpose, we enter marks p, q, or r into the squares, as in Fig. I.6. In one diagonal direction, the 11 squares from 15 to 51, from 36 to 63, and the squares 57 and 75 are marked p; the 11 squares from 25 to 52, from 37 to 73, and the squares 13 and 31 are marked q; the 11 squares from 35 to 53, from 47 to 74, and from 14 to 41 are marked r. Also, in the other diagonal direction, the 11 squares from 57 to 13, from 65 to 32, and the squares 73 and 51 receive a second mark p, the 11 squares 56 to 23, from 75 to 31, and the squares 37 and 15 receive a second mark q, the 11 squares 47 to 14, 55 to 33, and from 74 to 41 receive a second mark r.

		qq	rr	pp		
		pr	qp	rq		
pq	qr	rp	pq	qr	rp	pq
rr	pp	qq	rr	pp	qq	rr
qp	rq	pr	qp	rq	pr	qp
		rp	pq	qr		
		qq	rr	pp		

Fig. I.6.

These marks satisfy the same rules of addition as they did in section 3, namely

$$r \hat{+} p = q, \quad r \hat{+} q = p, \quad p \hat{+} q = r \tag{i}$$

$$r \hat{+} r = p \hat{+} p = q \hat{+} q = 0. \tag{ii}$$

This scheme is equivalent to that given by de Bruijn (1972). The marks of either set in three successive squares, horizontally or vertically, add up to 0. The marks of all squares of the board add up to (0,0).

By a move two squares are emptied, and one becomes covered. Whichever three squares are involved in this manner, the marks of either type in the former two squares add up to the mark of the latter square. We say that a configuration in which the marks of the two types add up to A and B respectively is in class (A,B). A configuration in any class can only be transformed into one in the same class.

The possible values of A and of B are $0, p, q, r$. Therefore there are 16 different classes. It is easy to find a representative configuration for any one of these classes. It is also easy to find two configurations such that no transformation is possible from one into the other, though they are both in the same class: the condition given for possible transformations is necessary, but not sufficient.

It is useful to know some sequences of moves which speed up the removal of blocks of counters, leaving the surrounding squares unaffected. Combination A removes three pegs in a row (Fig. I.7(A)). Combinations B_1 and B_2 remove two adjacent parallel rows of three pegs (Fig. I.7(B)) and combination C removes an L-formation of six pegs (Fig. I.7(C)).

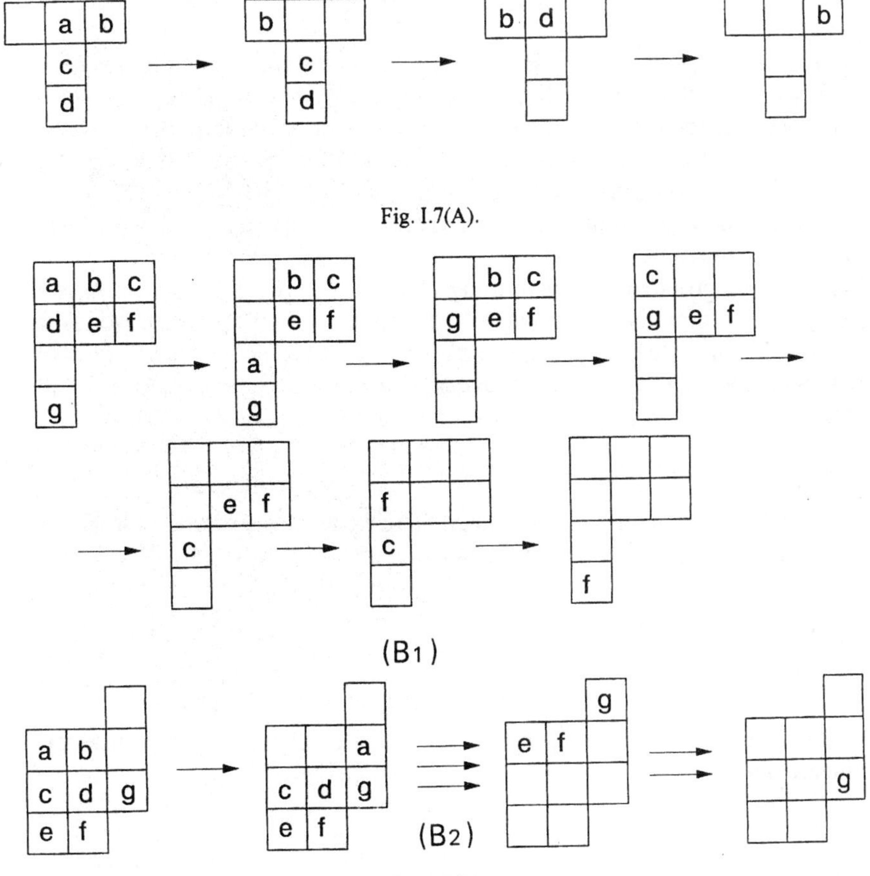

Fig. I.7(A).

(B1)

(B2)

Fig. I.7(B).

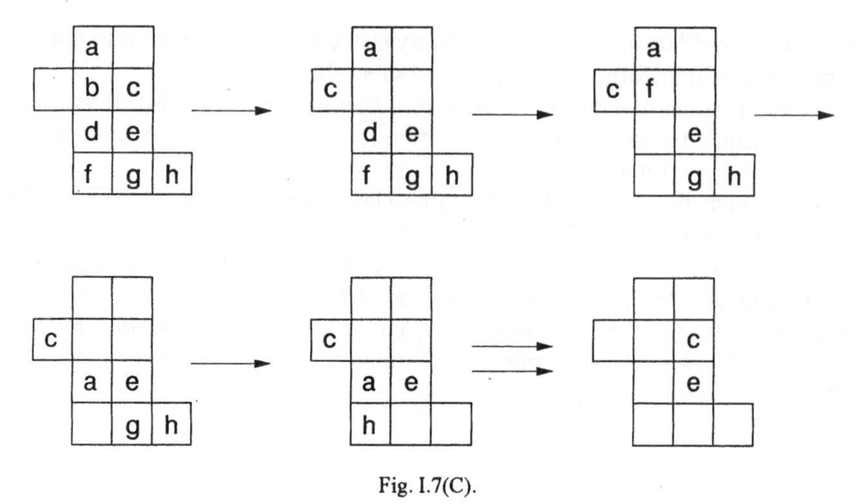

Fig. I.7(C).

Other useful moves are given in Beasley (1985), who credits some of them to Wiegleb (1779), which contains the earliest examples of play known to him. Beasley (1985) gives an encyclopedic survey of what is known about Peg Solitaire.

We add a few more remarks about the possibility of transformations. Denote a configuration where a given number of squares are empty, while the others are covered, by C, and its 'conjugate' transformation, where those empty in C are covered, and those covered in C are empty, by $\bar{C}$.

We write the problem of transforming C_1 into $\bar{C}_2$ as $[C_1–\bar{C}_2]$. Also, if M is a move which covers a formerly empty square and empties two squares, then we call a move which would empty the former square and cover the two latter squares (such a move is not legitimate), the 'conjugate' move M'. These concepts are used in the following theorem

Theorem A
If there exists a solution for $[C_1–\bar{C}_2]$, then there exists also a solution for $[C_2–\bar{C}_1]$, and the moves in the two solutions are the same, in reversed order.

Proof. Let the moves from C_1 to $\bar{C}_2$ be

$$M_1, M_2, \dots, M_n.$$

Conversely, if we start from $\bar{C}_2$ and want to reach C_1, then we might contemplate the conjugate moves, thus

$$M_n', M_{n-1}', \dots, M_1'.$$

Compare this with the sequence

$$M_n, M_{n-1}, \dots, M_1.$$

These are legitimate moves, and they transform C_2 into $\bar{C}_1$.

If C_1 and C_2 are identical, as in Fig. I.8, then the path from C_1 to $\bar{C}_2$ can be reversed, thus providing a second solution. The two solutions can not be identical. If the path consists of $2n + 1$ moves, then the nth and the $(n + 2)$th moves, which on the reversed path would be the $(n + 2)$th and nth moves respectively, would have to be the same, which is plainly impossible. If the path consists of $2n$ moves, the same argument applies to the nth and the $(n + 1)$th moves.

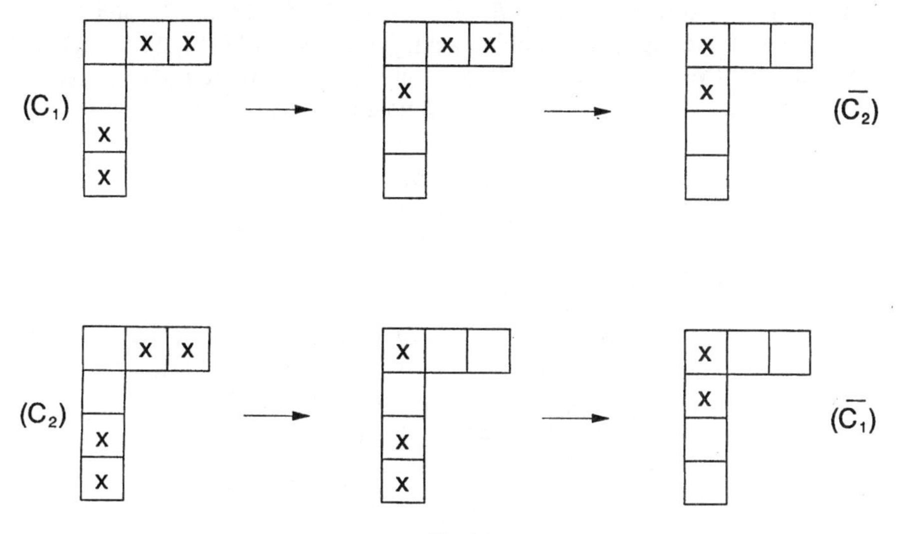

Fig. I.8.

Theorem B

Consider four squares *a, b, c, d,* horizontally or vertically adjacent, *a* empty and the other three squares covered. The jump *c–a* leaves *a* and *d* covered, but *b* and *c* empty. Precisely this same situation emerges if we start with only *d* empty, and *a, b,* and *c* covered, and perform the jump *b–d*. Therefore, if we know a succession of jumps from only *a* empty to some configuration, then this same configuration can also be reached by a path from only *d* empty.

It is also clear that if we can solve a problem P, then we can also solve a problem which derives from P by turning the original, and the final desired configuration through 90, 180, or 270°, clockwise or anticlockwise.

We can collect all problems whose solution is indicated by the solution of one of them into a family. Reiss (1857) has studied problems where originally only one square is empty, and finally only one square covered, and has shown that there are nine such families, which we identify by these two squares, namely

(44,44) (47,74) (47, 41) (45,45) (45,42) (42,42) (75,15) (55,55) and (25,52).

For instance, a solution of (44,44), starting with 64–44, indicates also the solution of (74,44) by Theorem B, and then of (44,74) by Theorem A. The solution of (47,74) gives us a clue to solving (74,47) (by Theorem A) and also solving (74,41), (41,14), (14,47), by clockwise turning. Reiss (1857) gives solutions to all nine families.

We look now at the family (44,44), which contains the standard problem: to begin with the central square is the only empty one, eventually it is the only covered one. The original configuration is in class (r,r). A possible solution of the standard problem (in fact, different from the one given by Reiss) is as follows:

Starting with 64–44, remove, using combination A, 52, 53, 54, and then 35, 45, 55. This leaves Fig. I.9. Now combination B_1 removes 63, 73, 64, 74, 65, 75, and 36, 37, 46, 47, 56, 57, to leave Fig. I.10. Combination B_2 removes 15, 25, 14, 24, 13, 23, and combination C removes 34, 33, 32, 31, 41, 51. Only two counters are still on the board, 42 and 43. 42 jumps over 43 into 44, and the standard problem is solved.

		37	47	57		
		36	46	56		
15	25				65	75
14	24	34			64	74
13	23	33	43		63	73
		32	42			
		31	41	51		

Fig. I.9.

15	25					
14	24	34				
13	23	33	43			
		32	42			
		31	41	51		

Fig. I.10.

The standard problem is one of those where, using the notation in Theorem A, (C_1) equals (C_2). As we have remarked, there are at least two different solutions for it. In fact, there are more than two.

Altogether 32 counters must be removed during the play. If we count a succession of moves by the same counter as one move, then we can ask for a solution with the smallest number of moves. The following solution, with 18 such moves, is due to Bergholt (1920).

> 46–44, 65–45, 57–55, 54–56, 52–54, 73–53, 43–63, 75–73–53, 35–55, 15–35, 23–43–63–65–45–25, 31–33, 34–32, 51–31–33, 13–15–35, 37–57–55–53, 36–43–32–52–54–34, 24–44.

Beasley (1962) has shown that no shorter solution exists. The proof is reproduced in Beasley (1985), pp. 135–139.

We introduce now another system of marking the squares of the board, which can be superimposed on any rectangular board, however large.

$$\begin{array}{ccccccccc} \cdots & \sigma^3 & \sigma^2 & \sigma & 1 & \sigma & \sigma^2 & \sigma^3 & \cdots \\ \cdots & \sigma^4 & \sigma^3 & \sigma^2 & \sigma & \sigma^2 & \sigma^3 & \sigma^4 & \cdots \\ \cdots & \sigma^5 & \sigma^4 & \sigma^3 & \sigma^2 & \sigma^3 & \sigma^4 & \sigma^5 & \cdots \\ \vdots & & & & & & & & \end{array}$$

σ is the positive solution of the difference equation

$$\sigma^i = \sigma^{i+1} + \sigma^{i+2}$$

that is $\sigma = \frac{1}{2}(\sqrt{5} - 1) \approx 0.618$. This value, and the pattern, are such that after a move the value in the square where the counter lands is either equal to, or smaller than, the sum of the values in the two squares which are emptied by the move. Beasley (1985) calls an assembly of values with this property a 'resource count', while J. H. Conway calls it a 'Pagoda Function' (see Berlekamp *et al.*, 1982, pp. 713–714). This property can be used to test the possibility of solving the following type of problem.

Draw a horizontal line across the board. All counters are, to begin with, below the line. The final configuration has one single counter somewhere above the line. (This has been called 'sending a scout ahead of an army'.)

It is easy to send a counter one step across the horizontal line, if there are two counters below it in appropriate places.

It is also easy to send a scout two steps ahead, if there are four counters suitably placed below the line

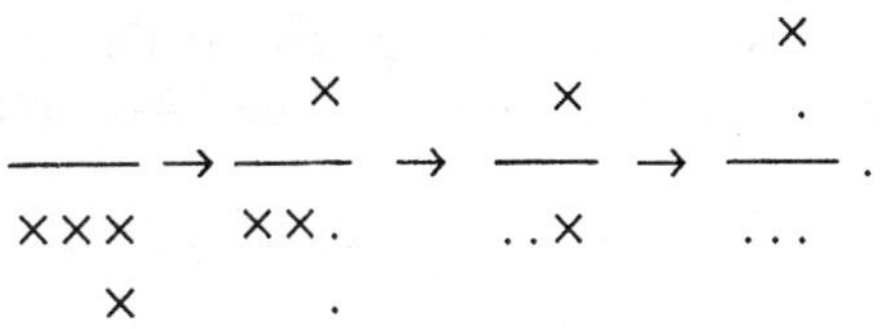

```
                                          ×
                  ×          ×            .
 ——————  →  ——————   →   ——————   →   ——————  .
 × × ×      × × .        . . ×        . . .
   ×          .
```

Now let us try to send a scout three steps ahead, into square 47 in class (r, r), when the horizontal line is drawn just above the squares 14, 24,...,64, 74, and the starting configuration consists of counters on squares 24, 34, 54, 64, 43, 42. This configuration is also in class (r, r), and there is no reason why we should doubt the possibility, using the criterion which we have developed.

However, let us put our pagoda pattern over the board, with mark 1 on the square 47, which the scout ought to reach. Compute the sum of the marks on the original counters:

$$\sigma^5 + \sigma^4 + \sigma^4 + \sigma^5 + \sigma^4 + \sigma^5 = 3\sigma^3 \approx 0.708,$$

which is smaller than 1. But every move reduces the total from originally 0.708, and can therefore not reach 1. Our problem can not be solved, unless we make use of more counters.

If, instead of just six counters, we make use of eight counters, on squares 24, 34, 44, 54, 64 and on squares 41, 42 and 43, then the problem can be solved in seven moves:

```
                                                                ×
                    ×              ×                ×           ·
 ______  →    ______    →    ______    →     _____    →    _____
 ×××××        ××·××          ··×××           ×××            ·××
   ×            ·              ·              ×              ×
   ×    43–45   ×     24–44    ×     41–43    ·     44–46
   ×            ×              ×              ·


                              ×
            ×          ×      ·
            ·          ×      ·
 →  ______   →  ______   →  ______
            ×··
            ×
 64–46         43–45         45–47
```

To move a scout four steps ahead, we need 20 counters to begin with. Berlekamp *et al.* (1982) gives two suitable formations, and Beasley (1989) has

```
_______
×××××××
××××××
×××××
  × ×
```

We leave it to the reader to puzzle out how to do it, and proceed to the question of sending a scout five steps ahead. Remarkably, this turns out to be impossible with a finite number of counters, however large.

Consider an infinite board. Put the mark 1 over the square we wish to reach. We have then just below the horizontal line

$$\ldots\sigma^8\,\sigma^7\,\sigma^6\,\sigma^5\,\sigma^6\,\sigma^7\,\sigma^8\ldots$$

$$\ldots\sigma^9\,\sigma^8\,\sigma^7\,\sigma^6\,\sigma^7\,\sigma^8\,\sigma^9\ldots$$

The total of all values in the row just below the line is

$$\sigma^5/(1-\sigma)+\sigma^6/(1-\sigma)=\sigma^4(1-\sigma)=\sigma^4/\sigma^2=\sigma^2$$

that in the next row further down is σ^3, and so on. The grand total of all the values below the line, infinite in number, is therefore

$$\sigma^2+\sigma^3+\ldots=\sigma^2/(1-\sigma)=1.$$

We need an infinite army to sent the scout ahead, and no finite army would do.

Observe that in the cases solved only moves were used which left the total of the pagoda function unchanged.

To complete this section, we add a few remarks about boards other than the English one.

The French board has in addition to the 33 squares of the English board four more squares, in positions we would call

22 (*qr*) 26 (*rq*)
62 (*rp*) 66 (*pr*).

If all squares are covered, the configuration is in class (*r,r*), and the configuration with only the central square empty is therefore in class (0,0). It is therefore impossible to solve the standard problem on the French board.

As an example which can be solved on this board, we quote from Lucas (1960, p.116) who refers to the *Encyclopédie méthodique. Dictionnaire des jeux mathématiques*, Paris an vii, p. 202 as the source of the problem:

Start: only square 44 empty
Finish: Fig. I.11

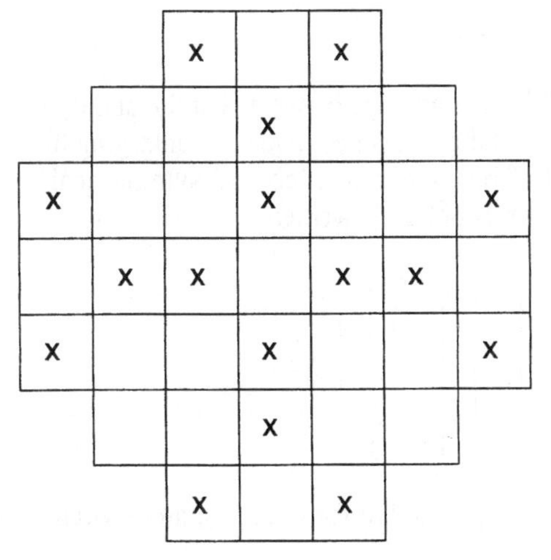

Fig. I.11.

Solution:

remove 25, 35, 45 by combination A, 47–45, 26–46
remove 32, 33, 34 by combination A, 14–34, 22–24
remove 43, 53, 63 by combination A, 41–43, 62–42
remove 54, 55, 56 by combination A, 74–54, 66–64 .

Reiss (1857), who developed more than a century ago many of the fundamental concepts which we have exhibited, mentions also a board with 41 squares, which contains in addition to those of the French board squares 04 (*qq*), 40 (*pp*), 48 (*qp*) and 84 (*pp*). No new concepts are involved when dealing with these boards.

5. MERLIN'S MAGIC SQUARE

Let a graph of n vertices be given. Each vertex has a 'parity', 0 or 1. A move means switching a vertex, that is changing the parity. When this happens, then every vertex connected by an edge to that which is being switched changes its parity as well.

Given some pattern of parities, it is required to change it by successive moves into another given pattern. For instance, start with all parities being 0, and change all of them into 1. A toy for playing this solitaire game has been commercially produced (cf. D. H. Pelletier, 1987).

When a vertex has been affected, either by its own switch, or by that of an adjacent vertex, an even number of times, then it resumes its original parity; if it has been affected an odd number of times, its parity will differ from its original one. For this reason it will be convenient if we perform all additions modulo 2. We shall denote this by the sign $\hat{+}$.

Let the original pattern of parities be $(x_1,...,x_n)$, and the pattern to be constructed be $(y_1,...,y_n)$. The play then has to change the parity of vertex k by $x_k \hat{+} y_k$. If the vertices $s, t, ..., u$ are switched on, in any order, then the parity of the vertex k changes by

$$a_{ks}x_s \hat{+} a_{kt}x_t \hat{+} \dots \hat{+} a_{ku}x_u$$

where a_{kj} equals 1, if there is an edge between k and j, and 0 otherwise. We let a_{ii} equal 1, for any i, and construct the adjacency matrix A (enlarging the usual adjacency matrix by 1s in the diagonal). This matrix is, of course, symmetrical.

In matrix notation our problem is written

$$\begin{pmatrix} a_{11} \dots a_{1n} \\ a_{21} \dots a_{2n} \\ \dots \\ \dots \\ a_{n1} \dots a_{nn} \end{pmatrix} \times \begin{pmatrix} z_1 \\ z_2 \\ \\ \\ z_n \end{pmatrix} = \begin{pmatrix} x_1 \hat{+} y_1 \\ x_2 \hat{+} y_2 \\ \dots \\ \\ x_n \hat{+} y_n \end{pmatrix}$$

Given $(x_1,...,x_n)$ and $(y_1,...,y_n)$, solve for $(z_1,...,z_n)$, and switch those vertices for which $z_i = 1$.

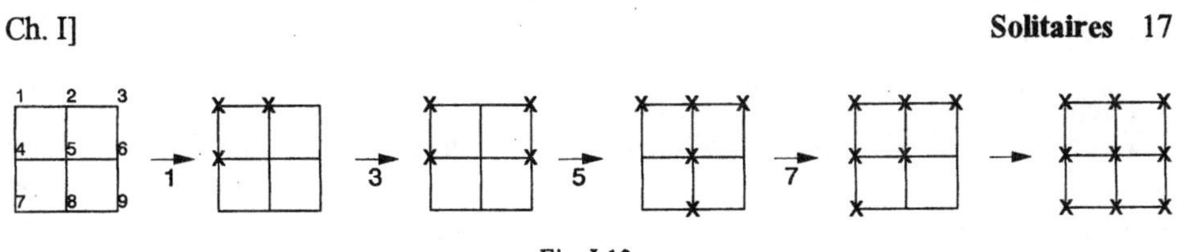

Fig. I.12.

Example 1

Consider the graph shown in Fig. I.12, start with all parities 0, and change all parities into 1. According to Suttner (1989), this is possible for any graph. We have now $x_i = 0$, all i, $y_j = 1$, all j. The matrix A equals

$$\begin{pmatrix} 1&1&0&1&0&0&0&0&0\\ 1&1&1&0&1&0&0&0&0\\ 0&1&1&0&0&1&0&0&0\\ 1&0&0&1&1&0&1&0&0\\ 0&1&0&1&1&1&0&1&0\\ 0&0&1&0&1&1&0&0&1\\ 0&0&0&1&0&0&1&1&0\\ 0&0&0&0&1&0&1&1&1\\ 0&0&0&0&0&1&1&1&1 \end{pmatrix}$$

and A^{-1} (all values modulo 2) equals

$$\begin{pmatrix} 1&0&1&0&0&1&1&1&0\\ 0&0&0&0&1&0&1&1&1\\ 1&0&1&1&0&0&0&1&1\\ 0&0&1&0&1&1&0&0&1\\ 0&1&0&1&1&1&0&1&0\\ 1&0&0&1&1&0&1&0&0\\ 1&1&0&0&0&1&1&0&1\\ 1&1&1&0&1&0&0&0&0\\ 0&1&1&1&0&0&1&0&1 \end{pmatrix}$$

Hence $(z_1,\dots,z_n) = (1\ 0\ 1\ 0\ 1\ 0\ 1\ 0\ 1)$.

Example II

On the graph of Example I, let

$$x = (1,0,0,1,1,0,1,1,0),\ y = (1,0,0,1,0,1,0,1,0).$$

so that

$$(x \hat{+} y)^{\mathrm{T}} = (0,0,0,0,1,1,1,0,0)^{\mathrm{T}}.$$

Therefore

$$A^{-1}(x \hat{+} y)^{\mathrm{T}} = (0,0,0,0,0,0,0,1,1)^{\mathrm{T}}$$

which means, switch on vertices 8 and 9, in any order.

Because , given the graph, and hence A and A^{-1}, the set $z_1,..,z_n$ depends only on $x_i + y_i$ $(i = 1,..., n)$, the vertices to be switched on are the same, whether $x_i = 1$ and $y_i = 0$ or $x_i = 0$ and $y_i = 1$. For instance, starting with $x_1 = 1$ and all other $x_i = 0$ and finishing with $y_1 = 0$ and all other $y_i = 1$, requires the same switching programme as that in Example 1.

If the matrix A is degenerate, so that no inverse exists, than a solution may exist, but may not be unique. We have such an example with a square with both diagonals, starting with all parities 0, to finish with all parities 1. Switching on any of the vertices solves the problem. On the other hand , (0, 0, 0, 0) into (1, 0, 0, 0) has obviously no solution at all.

6. INSTANT INSANITY

The game with this suggestive name uses four cubes with coloured faces. Altogether four colours are used, say Blue, Green, Red and Yellow. It is required to form the cubes into a column in such a way that each of the four sides of the column shows all four colours .

We describe a cube and its colours by four dots, marked B , G , R and Y. If one of the cubes sides is blue, say, and the opposite side is, say, green, then we connect the dots B and G by an edge, and similarly for the other pairs of opposite sides. If the cube has opposite sides of the same colour, we draw a loop at the corresponding dot. Each graph of a cube has thus three edges, a straight line or a loop .

A set commercially available will serve as an example. The four graphs are as in Fig. I.13(a).

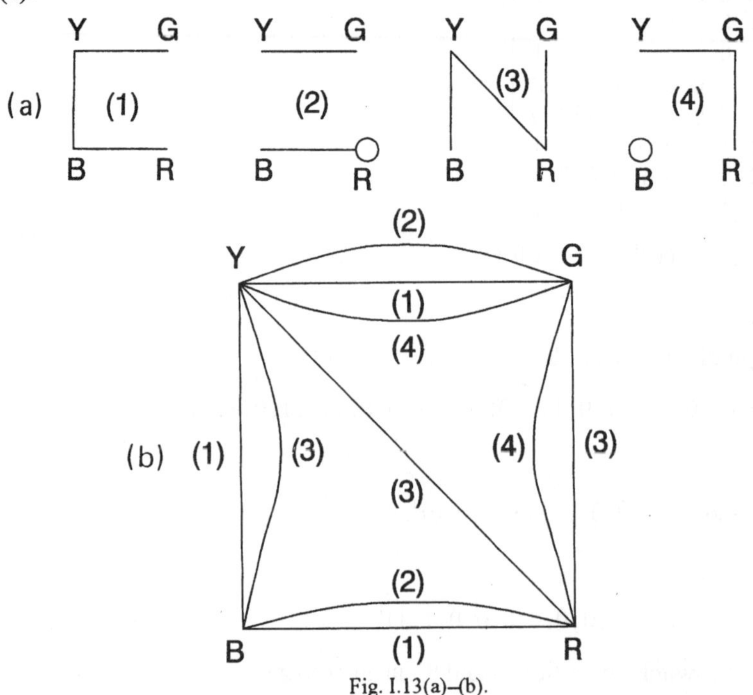

Fig. I.13(a)–(b).

Now look for a circuit in Fig. I.13(b) , where the four graphs of I.13(a) were superimposed. The circuit should have its four edges labelled (1), (2), (3) and (4), such as in Fig. I.14(a) .

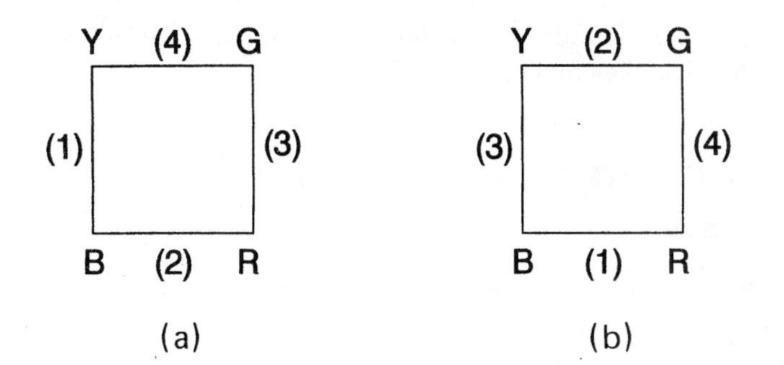

Fig. I.14(a)–(b).

This shows how to build a column such that the left-hand side as well as the right-hand side exhibits all four columns. Opened up, the column looks as Fig. I.15(a) .

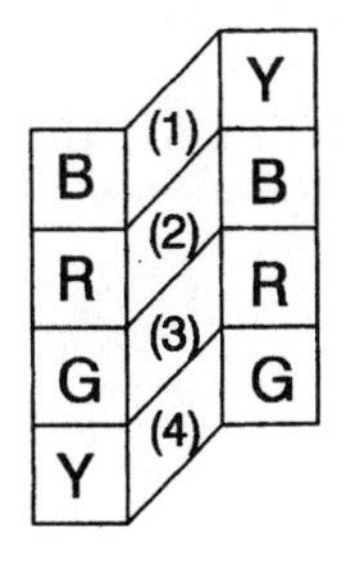

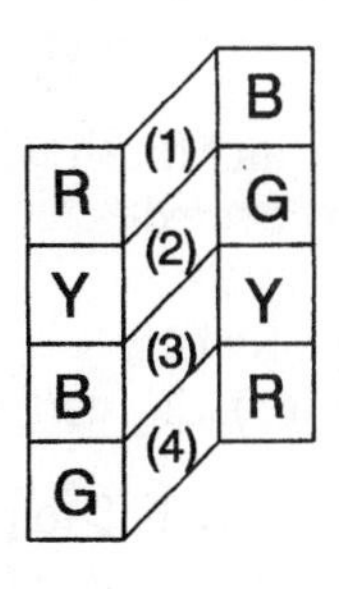

(a) (b)

Fig. I.15(a)–(b).

We must also achieve the desired effect on the front and on the rear sides of the column. To do this, we notice that the pattern so far obtained is not spoilt if a cube is rotated around the axis which joins the centres of the right- and of the left-hand sides. How do we find the appropriate rotations? We find them by looking at the second circuit which does not contain any edge of the first, for instance that in Fig. I.14(b). Again, this circuit tells us which colour should appear, on any one of the four cubes, in front and in the rear. If we turn the cubes accordingly, we obtain in our example the column, opened up, shown in Fig. I.15(b). We have thus solved our problem.

According to Tucker (1980), the method shown is due to F. de Carteblanche, reputedly no other than the graph theorist W. Tutte.

EXERCISES TO CHAPTER I

1. The Fifteen Puzzle

(a) Change into

1	15	9	5		1	15	9	5
7	14	3	11		7	2	14	3
12	2	10	6		12	13	10	11
4	13	8	.		4	8	6	.

(b) Change into

1	15	9	5		1	15	9	5
7	14	3	11		7	2	14	3
12	2	10	6		12	13	11	10
4	13	8	.		4	8	6	.

2. Triangular Solitaire

Start with only triangle 3 empty (Fig. I.4), and transform into only triangle 2 covered.

3. Peg Solitaire

(a) On the English board, start with only square 44 empty, and finish with only the squares 35, 55, 53, 33 covered.

(b) Start as in (a), and finish with only the squares 13, 14, 15, 31, 41, 51, 73, 74, 75, 37, 47, 57 and 44 covered.

4. Merlin's Magic Square

On the graph of Fig. I.12, start with all parities 0 and

(a) change vertices 2, 4, 6, 8 to parity 1.
(b) change vertices 1, 3, 5, 7 9 to parity 1
(c1) change vertices 2, 4, 5, 6, 8 to parity 1
(c2) change vertex 5 to parity 1.
(d1) change vertices 1, 2, 4, 6, 8, 9 to parity 1
(d2) change vertices 1, 9 to parity 1.

II

Games for two players

In this chapter we discuss games in which two players move alternately, according to well-defined rules. Only a finite number of moves are possible from any position, and the game ends with either one of the players winning, or it ends in a draw, after a finite number of moves. Moreover, there are no chance moves in our games, and at each stage both players know all the previous moves of either player ('perfect-information' games).

A. TAKE-AWAY GAMES

These games are played with one or more piles of counters. Each player reduces the number of counters and the player who makes the last possible legal move wins.

To deal with these games, we introduce a function of the position, that is of the sizes of the piles. Grundy (1939) introduced this function when analysing the game of Nim (see later). Sprague (1935–6) had in fact mentioned the idea of such a function earlier, and we shall refer to the function as the Sprague–Grundy numbers, or the Grundy function, abbreviated to the g-numbers, or g-function.

The Grundy function is constructed recursively in the folliwng way. A 'terminal' position, that is a position from which no further move is possible — either by the rules, or because there is no counter left — has the g-number 0. To obtain the g-number for any other position, consider the set of all positions which can be reached by a move, and their g-numbers. The Grundy number of the position to be marked is then the smallest non-negative integer not contained in the set of those g-numbers of follower positions. This minimum exempted value is sometimes called mex in the literature. The usefulness of the Grundy function derives from its properties, which we shall now exhibit.

Clearly, no position with g-value v can have a follower with the same value. On the other hand, it has some follower with any smaller g-value, unless v equals 0.

It follows, that if a player produces by his move a position with value 0, his opponent must move, if he can move at all, to a position with positive value. The first player can then produce once more a position with value 0. Eventually, he will be able to reach a terminal position (with value 0) and win.

Observe, though, that a player can not necessarily move from a position with positive value to any position with value 0.

Because of the property of positions with g-value 0 we call them *winning* positions, that is winning for the player who produces it. However, if the play starts from such a position, then it is the opponent of the player whose turn it is, who has the advantage.

Admittedly, when a winning strategy, for the first or for the second player, is known, then actually playing the game loses all interest. But for us it is not the play which is the thing (*Hamlet*, II, ii), but its theory.

1. SUBTRACTION GAMES

These are games played with one pile of counters. A *subtraction set* $s_1, s_2, \ldots$, $0 < s_1 < s_2 < \ldots$ is given, and a legal move consists of removing s_i counters from the pile. The player who leaves less than s_1 counters in the pile wins.

We denote this game by $S(s_1, s_2, \ldots)$, and the g-value of a pile of n counters is called $g(n)$.

As an illustration, take the game $S(1,3,4)$. The sequence of g-values starts with

n	0	1	2	3	4	5	6	7	8	9	10	11	12
$g(n)$	0	1	0	1	2	3	2	0	1	0	1	2	3

If the subtraction set is finite, with k elements, then $g(n)$ is the smallest non-negative integer different from $g(n-s_1), \ldots, g(n-s_k)$. If these values are $0, 1, \ldots, k-1$, then $g(n) = k$; otherwise $g(n) < k$. It can be shown that the sequence $g(n)$ must therefore be periodic, though the period may be very large and unpredictable.

A known sequence of g-numbers does not define a substraction game uniquely. This follows from the following proposition:

If no $g(n)$ equals $g(n-s)$ for any n, then s can be added to the subtraction set, without changing the sequence of g-numbers. This is so, because $g(n-s)$ does not affect that set of g-values which define $g(n)$.

There exists a simple relationship between the $g(n)$ sequence of a game $S(s_1, s_2, \ldots)$ and another game $S(ts_1, ts_2, \ldots)$. The sequence of the second game is easily seen to derive from that of the first game by repeating every $g(n)$ t times. Consequently, if $s_1, s_2, \ldots$ have a greatest common divisor d, then the g-sequence can be found from that of $S(s_1/d, s_2/d, \ldots)$ by repeating d times each term of the sequence of the latter.

We consider now in more detail some subtraction games with a finite set $(s_1, s_2, \ldots, s_k)$.

The game $S(1)$ has obviously the g-sequence $\dot{0}\dot{1}$ (that is 010101...), and hence $S(t)$ has the sequence with period $\dot{0}\ldots01\ldots\dot{1}$, 0 as well as 1 repeated t times.

For $S(1, s_2)$ we must distinguish between odd s_2 and even s_2:

(a) s_2 is odd.
The sequence of $g(n)$ is periodic with period $\dot{0}\dot{1}$. As a consequence of the proposition above, the set $(1, s_2)$ can be extended by any odd s.

(b) s_2 is even.

$S(1,2)$. n 0 1 2 3 4 5 6 7 8 ...
$g(n)$ 0 1 2 0 1 2 0 1 2 ...

$S(1,4)$ n 0 1 2 3 4 5 6 7 8 9 ...
$g(n)$ 0 1 0 1 2 0 1 0 1 2 ...

and generally for $S(1,2t)$ we have a periodic sequence with period

$$\underbrace{\dot{0}\,1\,0\,1\ldots 0\,1}_{t\text{ pairs}}\ \dot{2}$$

Generalizing from $S(1,2)$, consider $S(1, 2,..., M)$.
This game has a $g(n)$-sequence with period $\dot{0}$ 1 2 ... $\dot{M}$.

The winning positions are those of n being a multiple of $M+1$. That much we could have found out without constructing any sequence. Clearly, the opponent can not construct another such pile, because he can not remove more than M counters. If he takes $c < M$, the first player takes $M + 1 - c$, thus recovering a winning position. Eventually he takes the whole pile.

Because of the periodicity of the g-sequence, the subtraction set can be extended by adding to it

$$2+M,\ 3+M,\ \ldots,2M+1$$
$$3+2M, 4+2M,\ldots,3M+1$$

and so on. Generalizing further, consider $S(m, m+1, ..., M)$. Its g-sequence is periodic with period

$$\underbrace{0\ldots0}_{m}\ \underbrace{1\ldots1}_{m}\ \ldots\ \underbrace{(t-1)..(t-1)}_{m}\ \underbrace{t..t}_{s}$$

times,

where t is the largest integer not exceeding $(M+m)/m$, and $s = M+m-mt$. The length of the period is $M + m$. For instance, $S(2, 3, 4, 5, 6, 7)$ has the period $\dot{0}$ 0 1 1 2 2 3 3 $\dot{4}$.

The player who produces a pile size congruent to 0, 1, 2, ..., $m-1$ modulo $M+m$ can win, because at each move he can reconstruct such a pile size. Eventually, he leaves 0, 1, 2, ..., or $m-1$ counters, and his opponent has no possible move to make.

Berlekamp *et al.* (1982) contains a list of g-sequences for all subtraction games with $s_i \leq 7$ $(i = 1,2,...,7)$.

Let us now look at subtraction games with an infinite subtraction set.

In the game $S(1,2,...)$ we have $g(n) = n$; the first player takes the whole pile and wins. This one-pile game is, of course, trivial, but we shall have to remember it when we deal with multi-pile games.

S *(all odd prime numbers)*

(a) 1 counts as a prime. This is simply the game $S(1, 3)$ extended, with period $\dot{0}\dot{1}$.

(b) 1 does not count as a prime.

n	0	1	2	3	4	5	6	7	8	9	10	11	12	13	14	15	16	17	18	19
$g(n)$	0	0	0	1	1	1	2	2	2	3	0	3	4	1	4	3	0	3	4	1

S *(all prime numbers)*

(a) 1 counts as a prime number. This is the game $S(1, 2, 3)$ extended, with period $\dot{0}12\dot{3}$.

(b) 1 does not count as a prime number.

n	0	1	2	3	4	5	6	7	8	9	10	11	12	13	14	15	16	17	18	19...
$g(n)$	0	0	1	1	2	2	3	3	4	0	0	1	1	2	2	3	3	4	4	5...

Note: this starts like the sequence for $S(2, 3, 4, 5, 6, 7)$, but differs from it after $n = 17$.

S *(all Fibonacci numbers* F_i*)*

$$F_1 = F_2 = 1, F_{i+2} = F_{i+1} + F_i .$$

n	0	1	2	3	4	5	6	7	8	9	10	11	12	13	14	15	16	17	18	19	20...
$g(n)$	0	1	2	3	0	1	2	3	4	5	0	1	2	3	0	1	2	3	4	5	0...

This looks like a periodic sequence with period 10, but that cannot be true, because if it were, then $g(34)$ would be 0. However, 34 happens to be a Fibonacci number, viz. F_8, so that a pile of 34 counters could be reduced to 0. Hence $g(34)$ cannot be 0; it is in fact 1.

We ought to mention that some authors, for instance Br. U. Alfred (1963) calls this subtraction game Fibonacci Nim. It is, however, different from the Whinihan (1963) game of the same name, which is not a subtraction game, because the legality of a move depends on previous moves. We shall deal with the latter game later.

S *(all Lucas numbers* L_i*)*

$$L_1 = 1, L_2 = 3, L_{i+2} = L_{i+1} + L_i .$$

n	0	1	2	3	4	5	6	7	8	9	10	11	12	13	14	15	16	17
$g(n)$	0	1	0	1	2	3	2	3	0	1	0	1	2	3	2	3	0	1...

To conclude that $g(n)$ has a period of length 8 would be wrong, because $g(18)$ equals 2, not 0. 18 is, in fact, L_6.

For subtraction games the following two theorems hold (Ferguson, 1974).

Theorem I

(a) If $g(n) = 0$, then $g(n + s_1) = 1$.

(b) if $g(n) = 1$, then $g(n - s_1) = 0$.

Observe that $(n - s_1)$ must be non-negative, because if n were smaller than s_1, the smallest of the s_i, then no move would be possible from n, and hence $g(n)$ would be 0, not 1.

Proofs.

(a) $g(n + s_1) \neq 0$, because if it were, $g(n)$ could not be 0. Now assume, by contradiction, that (a) does not hold for all n, and let x be the smallest value for which $g(x) = 0$, but $g(x + s_1) \neq 1$.

Then there must be some s_i $(>s_1)$ such that

$$g(x + s_1 - s_i) = 1. \qquad \text{(i)}$$

Consequently $x + s_1 - s_i$ must have a follower. This follower, obtained by subtracting s_1, or s_2,... will be $x - s_i$, or smaller. In any case, $x - s_i \geq 0$. Moreover, since $g(x) = 0$, we have $g(x - s_i) > 0$.

It follows, that there must exist some s_j such that

$$g(x - s_i - s_j) = 0. \qquad \text{(ii)}$$

(i) implies that $g(x + s_1 - s_i - s_j) \neq 1$. Write $y = x - s_i - s_j$, then $y < x$, $g(y) = 0$, $g(y + s_1) \neq 1$, which contradicts the definition of x. This proves (a).

(b) If $g(x) = 1$, but $g(x - s_1) \neq 0$, the there is some s_k such that $g(x - s_1 - s_k) = 0$, and it follows then from (a) that $g(x - s_k) = 1$, which would contradict $g(x) = 1$.

Theorem II

If $g(x) = 0$, and further moves are possible, then there is some s_i such that $g(x - s_i) = 1$.

Observe that this does not follow from the construction of the Grundy function $g(n)$. Indeed, $g(x) = 0$ could conceivably be true even if all pile sizes smaller than x had Grundy numbers at least as large as 2.

Proof. If $g(x) = 0$, then $g(x - s_1) \neq 0$, hence there is some s_i such that $g(x - s_1 - s_i) = 0$. Then it follows from Theorem I (a), that $g(x - s_i) = 1$.

Games have been proposed which might be called addition games $A(s_1,...,)$. Starting from 0, the two players add, turn by turn, any of the s_i to the sum already obtained, and the player who first reaches a given target wins. If the target cannot be precisely reached, then the first player who would, by adding any of the s_i, exceed the target, loses. This game is obviously, equivalent to $S(s_1,...)$ and its $g(n)$ sequence is the same as the sequence $g(T-n)$ for the subtraction game, T being the target. Thus, for instance the sequence for $A(1,3,4)$ with target 12 is

n	0	1	2	3	4	5	6	7	8	9	10	11	12	...
$g(n)$	3	2	1	0	1	0	2	3	2	1	0	1	0	...

2.

From subtraction games, we turn to other take-away games

Game A

Let it be allowed to remove a number of counters from a pile of size n, if the number to be removed is a square, and relatively prime to n.

n	0	1	2	3	4	5	6	7	8	9	10	...
$g(n)$	0	1	0	1	0	2	0	2	0	1	2	...

Game B

From a pile of size n, a number of counters may be removed if that number is a divisor of n.

n	0	1	2	3	4	5	6	7	8	9	10	11	12	...
$g(n)$	0	1	2	1	3	1	2	1	4	1	2	1	3	...

If n equals 2^{k-1}, then $g(n)$ equals k. (Proof by induction.)

Now consider games where a player does not only win when he reduces the pile size to 0, but also if he reduces it to t_1, or t_2,..., or t_k.

As a example, take $S(1,3,5)$ with the periodic g-sequence $\dot{0}\dot{1}$, and add the rule that the player who produces a pile size of 3 or of 5 also wins. To construct the g-function, insert 0 at positions 0, 3, and 5 to begin with, and compute the remaining positions in the ordinary way. This produces

n	0	1	2	3	4	5	6	7	8	9	10	11	12	13	14	15	...
$g(n)$	0	1	0	0	2	0	1	3	1	0	1	0	1	0	1	0	...

As a second example, take $S(1,2)$, which has periodic g-sequence $\dot{0}2\dot{1}$, and again $t_1 = 3$, $t_2 = 5$.

n	0	1	2	3	4	5	6	7	8	9	10	11	12	...
$g(n)$	0	1	2	0	1	0	2	1	0	2	1	0	2	...

For sufficienctly large n, the sequence becomes periodic with period $\dot{0}2\dot{1}$.

We shall now mention games where the legality of a take-away move depends not only on the position reached, but also on the history of previous moves.

Schuh's Game

As we have seen, in the subtraction game $S(1,2,...,M)$, the player who first produces a size equal to a multiple of $M+1$ can win. We make this game somewhat less transparent by adding a further condition: the number of counters taken away in two successive moves by the two players must not add up to $M+1$. The winning strategy mentioned is then out of order (Schuh, 1968).

The winning strategy aims now at producing pile sizes which are multiples of $M+2$. This is easy, unless the previous move has just removed one single counter, because then you would have to remove $M+1$ counters, which is not allowed. In that case you remove also 1, which produces $t(M+2)-2$; the opponent cannot produce now a winning position, because to do this, he would have to take away M counters, which is forbidden by the additional rule. He will have to produce a position from which you can regain a winning one.

Example

$M=7$. No addition to 8. Winning positions $9t$.

$$\begin{array}{l} 19 \to 18 \to 12 \to 9 \to 8 \to 7 \to 5 \to 0\text{. A wins.} \\ \quad\text{A} \quad \text{B} \quad \text{A} \quad \text{B} \quad \text{A} \quad \text{B} \quad \text{A} \end{array}$$

In another game, also due to Schuh, the additional rule mentioned above is replaced by the rule that no two successive removals must be equal in numbers. In this case, if M is even, hence $M+1$ is odd, the first player can always reconstruct a winning position $t(M+1)$, because no sum of equal numbers can be $M+1$ in any case. But for odd M no such general rule exists.

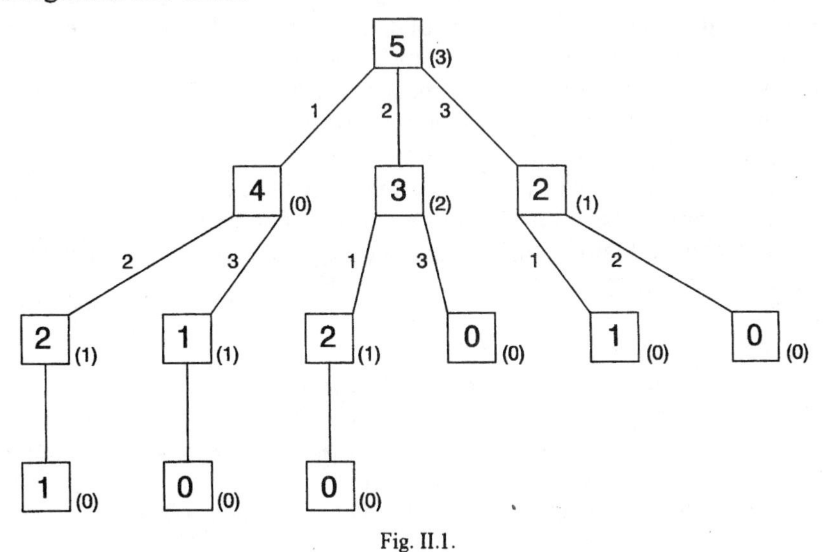

Fig. II.1.

It is still possible to attach g-numbers to any position, provided a position is defined not merely by the pile size, but also by the relevant pattern of the way in which it was reached. For example, let the starting pile size be 5 and $M=3$. In Fig. II.1, the pile sizes are shown in squares, and the g-numbers in round brackets. It will be noticed, that the pile size 2 appears twice in the third row. The previous histories of these positions are different, and the possible next moves differ as well.

Schuh (1968) discusses such cases extensively.

Binary Nim

We have seen that the subtraction game $S(1,2,...)$ is trivial. But let us introduce the further rule that no player may start the play by removing the whole pile, and that no player may remove more counters than his opponent has done at his last move. The last player able to move wins (Schwenk, 1970).

It would obviously be a mistake for any player to take half or more than half of the counters in the pile; the opponent would take the whole pile and win.

For the argument which follows it is convenient to think of the pile size to be expressed in binary notation, that is as a sum of powers of 2.

Assume, to begin with, that the size of the pile is not a power of 2, so that the number of digits 1 in the notation is larger than 1. The following strategy is then winning. Remove the largest power of two which divides the pile size. This largest power is indicated by the rightmost digit 1 in the notation.

Proof. If the pile size is odd, that largest power is 2^0, i.e. 1. This is taken by the first player. All later moves by either player will again be the removal of 1, and the first player will take the last 1 and win.

If the pile size is even, the first player takes the number indicated by the right most digit in the notation. Thereby the number of digits 1 is reduced by 1. The second player cannot use the winning strategy, because this would mean removing at least twice the previous amount. He takes less, and his move will not reduce the number of 1s in the notation, but will produce a digit 1 more to the right. This enables the first player to remove again the largest power of 2 dividing the pile size.

Eventually the size will reduce to zero, and this cannot be the effect of a move by the second player, because he never reduces the number of digits 1.

On the other hand, if the original pile size is a power of 2, the second player can win by using the winning strategy.

Fibonacci Nim

This game (Whinihan, 1963) is played according to these rules:

The first player may not remove the whole pile at his first move. No player may remove more than twice the number of counters which his opponent has removed at the last move. The last player who can move wins.

Clearly, no player should ever take one-third or more of the remaining counters in the pile.

Provided the pile size is not a Fibonacci number, the first player wins with the following strategy (for the explanation of the concepts used, see Appedix III):

— When allowed, take the whole pile.
— Otherwise express the pile size canonically as a sum of Fibonacci numbers, and take away the number of the smallest term in this expression.

It follows from Property I (see Appendix III) that the opponent cannot use now the winning strategy; he must take a smaller number than the smallest term in the canonical representation of the pile size facing him. By Property II (Appendix III) the first player can use once more the winning strategy, and he will eventually win.

On the other hand, if the initial pile size is a Fibonacci number, then the first player cannot prevent the second player from using the winning method.

In the last two games which we have mentioned the moves were conditioned by a rule which restricted the number of counters to be taken away by a multiple of the number removed at the previous turn. The multiplying factor was 1 in Binary Nim, and 2 in the Fibonacci Nim game. In the next game we generalize this rule (Schwenk, 1970).

Schwenk's Nim

The number of counters removed may not exceed some given function of the number removed at the opponent's last move. This function, $f(n)$, must satisfy the conditions

$$f(n) \geq n, \text{ and } f(n) \geq f(n-1).$$

These conditions are indeed satisfied in the cases of Binary Nim and Fibonacci Nim.

To find the winning strategy, define a sequence H_i by the recursion

$$H_1 = 1, H_{k+1} = H_k + H_j,$$

where j is the smallest subscript for which $f(H_j) \geq H_k$.

Such a j exists, because $f(H_j) \geq H_j$.

When $f(H_j) = H_k$, then $H_2 = H_1 + H_1$, $H_3 = H_2 + H_2$; generally $H_i = 2^{i-1}$.

When $f(H_j) = 2H_j$, then $H_2 = H_1 + H_1 = 2 = F_3$, $H_3 = H_2 + H_1 = 3 = F_4$, generally $H_{n+1} = H_n + H_{n-1} = F_{n-2}$.

The H_i are, in this case, Fibonacci numbers.

Schwenk (1970) has also proved that any positive integer N can be expressed uniquely in the form

$$N = \sum_{i=1}^{n} H_{j_i}, \text{ with } f(H_{j_{i+1}}) < H_{j_i}. \qquad (*)$$

This is the equivalent of Zeckendorf's theorem for Fibonacci numbers and can be proved in an analogous manner. The condition (*) is Property I in the Fibonacci case and a property corresponding to Property II in the latter case holds also generally. Consequently, the winning strategy consists of expressing the pile size as such a sum of H-numbers, and removing the smallest H which appears in the sum.

We give an example (Fig. II.2) for the case $f(n) = 3n$, where the sequence of H_n starts with

n	1	2	3	4	5	6	7	8	9	10
H_n	1	2	3	4	6	8	11	15	21	29

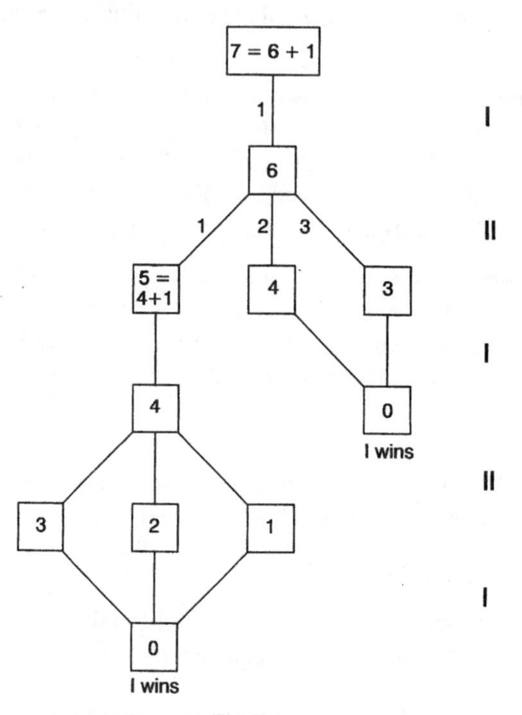

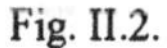
Fig. II.2.

In all those games which we have described so far, the possibilities or usefulness of moves did not depend on which player's move it was. Such games are called 'impartial'. Now we introduce games which are not impartial; they are 'partizan' games. To set the scene, we exhibit an otherwise trivial game.

Trivial Game

The players are E, who is allowed to take away an even number of counters, and O, who takes an odd number. The player who has the first move will be called A, the other B.

First, let the initial size be even. Then, if A is E, he simply takes the whole pile. But if A is O, then hc lcaves a single counter, which E cannot take and therefore loses.

On the other hand, start with an odd size. Then O wins in any case. If he is A, he takes all counters, if he is B, then E, who is A, must have left after his first move an odd number of counters, for O to take and so to win.

More complex games of this type are considered by Maulden (1978).

To turn to a much more interesting, but not yet completely explored, scenario, consider games which fail to be impartial because the final result depends on all previous moves of the two players.

Odd Wins

We start with an odd pile size n. Either player can take up to k counters. The player who has, in all his moves, taken an odd total number of counters wins.

It is easy to see that if $k = n - 2$, the second player can win. If the first player takes an odd number of counters, say s, and hence leaves an even number, then B takes one less than the whole pile, leaving 1 which the first player must take, to lose. If A takes an even number, leaving an odd one, B takes all the rest.

There are, in the literature, examples of cases where k is not $n - 2$, such as Dudeney (1917, Problem 392), or Sprague (1963), but no general solution is known (at least to the present author).

O'Beirne's Games

Each player takes 1, 2, or 3 counters from the pile and keeps them. At the end of the game he discards a multiple of three of the counters he has accumulated. The player who has more counters left wins. If the numbers left are equal, then the result is a draw. This time we cannot indicate a winning strategy, but only one which leads to a win or a draw.

If the initial pile size is $3t$, take first 2, and afterwards 3 as long as possible. If this is possible throughout, then you will have taken $3s + 2$ counters, and the opponent $2(t - s - 1) + 1$: you win. However, if towards the end of play you are faced with a pile of size 1 or 2, take 1. The result will be 0 against 0, a draw.

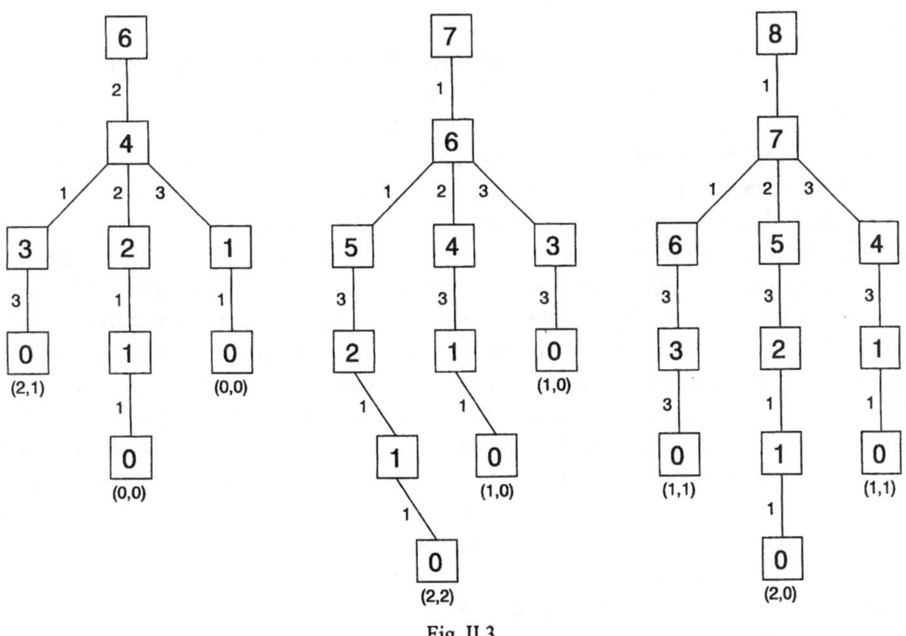

Fig. II.3.

If the initial size is of the form $3t + 1$, take first 1, then 3 as long as you can. If this is possible throughout, you will have taken $3s + 1$, your opponent will have taken $3(t - s)$. Once more, you win. But if you are faced with 1 or 2 counters, take 1. This results in 2 against 2, a draw.

If the initial size is $3t + 2$, take 1, then 3. If you can take 3 up to the end of the play, you will have taken $3s + 1$, your opponent $3(t - s) + 1$. The result is 1 against 1, a draw. But if, towards the end, you are faced with 1 or 2, take 1. You win 2 against 0.

The best the second player can do is to force a draw. To achieve this, he repeats your moves as long as he can, then takes 1.

The cases $3t$ and $3t + 2$ appear to have been invented by O'Beirne (1965).

We illustrate the three cases in Fig. II.3. Only those edges are drawn for the first player which form part of his strategy explained above.

3.

All the games so far discussed were based on counters in one single pile. Now we shall discuss games with more than one pile while we shall use the theory of one-pile games as building blocks.

Nim

The simplest multi-pile game, and the prototype for many other games, is the game of Nim. The origin of the name is obscure, but in any case irrelevant to our purpose.

It is played with any number of piles, of any sizes. Any positive number of counters may be removed by the two players in turn, all of the counters from the same pile. The player who takes the last counter wins. This is, of course, an impartial game. (Actually, the more usual rule stipulates that it is the player who must take the last counter loses. We shall discuss this mathematically more complex version later, as a misère game.) The winning strategy, which we now describe, was discovered by Bouton (1902).

Write the pile sizes in binary notation. Then add the numbers referring to the same power of 2, modulo 2, that is call the sum 1 if the answer is odd, and 0 if it is even. The total is again interpreted in binary notation. We call this addition 'without carry' Nim-addition, and denote it by a plus sign with a French circumflex accent, thus $\hat{+}$. Nim-addition is commutative, but adding a number to another by Nim-addition need not increase either number. In particular, the Nim-sum of two equal numbers is zero.

The resulting Nim-sum of the sizes of a set of piles is called the g-number of the configuration of pile sizes. The winning strategy consists of producing a Nim-sum 0, that is a sum whose binary digits are all 0.

In order to prove that this strategy is winning, we must prove that

(a) it is impossible to move from a winning position, that is one with Nim-sum 0, to another winning position, and
(b) it is possible to move from any position which is not winning to one which is winning.

To justify the name g-number for the Nim-sum described we prove that

(c) the Nim-sum is the smallest non-negative integer not in the set of the Nim-sums of all those positions which can be reached in one move (all *follower* positions). In other words, any smaller Nim-sum can be generated by some move, not just 0, provided the present Nim-sum is not 0.

Proofs.

(a) This is immediate. Any change of one of the summands — any move — changes at least one of the digits from 0 to 1.

(b) Find (one of) the pile size(s) with a left-most digit 1 at the same highest rank as that in the sum. Add the sum to that pile number by Nim-addition. This reduces the pile size and — since Nim-addition is commutative — adds the sum to itself, producing 0.

(c) To change the sum from S into T, say, add $S \hat{+} T$ to S (remember $S \hat{+} S = 0$). Find (one of) the pile size(s) with a digit 1 at the same rank as the left most digit 1 in $S \hat{+} T$. Add $S \hat{+} T$ to the pile size, which is thereby reduced, as required. If T equals 0, this reduces to the procedure described in (b) above.

Examples

Piles 3, 5, 8.

In binary notation 3 ~ 11
5 ~ 101
8 ~ 1000
Nim–sum 1110 ~ 14.

Procedure (b) Add 1110 to 1000, sum 110. Reduce 8 to 6.

Result 3 ~ 11
5 ~ 101
6 ~ 110
000

Procedure (c) $T = 10 \sim 1010$. Add 1110 to 1010, sum 100.
Add 100 to 101, sum 1.

Result 3 ~ 11
1 ~ 1
8 ~ 1000
1010.

Unless the play starts with a g-sum 0, the first player will win by moving at each turn to a winning position, with g-number 0.

It is worth looking at the special case of two-pile Nim.

The Nim-sum of two numbers is zero, if and only if the two numbers are equal. If the two pile sizes differ, reduce that of the larger one to that of the smaller pile. The opponent cannot make them equal again but the first player can do it at his turn. Eventually he will win. However, if the pile sizes are equal to begin with, the second player has the winning strategy.

This winning strategy can be described, fancifully perhaps, by saying: first, make the two piles of equal size; afterwards *echo* the move of the opponent. If he takes c counters, you take c counters from the other pile, to make the sizes equal again. This echoing method, repeating a move in another portion of the position, will be found useful in a number of games to be described.

It follows from the fact that the Nim-sum of a two-pile Nim is the g-number of the position, that a Nim-addition table can easily be constructed. The entry in row x and column y, that is $x \dotplus y$, is the smallest non-negative number which is not in the set of the entries for smaller x and the same y, (on the left of the entry) or of the entries for smaller y and the same x (above the entry). We obtain the following table.

$x=$	0	1	2	3	4	5	6	7	8	9	10	11	12
$y=$ 0	0	1	2	3	4	5	6	7	8	9	10	11	12
1	1	0	3	2	5	4	7	6	9	8	11	10	13
2	2	3	0	1	6	7	4	5	10	11	8	9	14
3	3	2	1	0	7	6	5	4	11	10	9	8	15
4	4	5	6	7	0	1	2	3	12	13	14	15	8
5	5	4	7	6	1	0	3	2	13	12	15	14	9
6	6	7	4	5	2	3	0	1	14	15	12	13	10
7	7	6	5	4	3	2	1	0	15	14	13	12	11
8	8	9	10	11	12	13	14	15	0	1	2	3	4
9	9	8	11	10	13	12	15	14	1	0	3	2	5
10	10	11	8	9	14	15	12	13	2	3	0	1	6
11	11	10	9	8	15	14	13	12	3	2	1	0	7
12	12	13	14	15	8	9	10	11	4	5	6	7	0

This method of constructing a Nim-addition table can be generalized to Nim with more than two piles. For the addition of three numbers, each entry is the smallest non-negative number not in set of those above it in a higher level square, to the left of it, or above it in its own square.

0

	0	1	2	3	4
0	0	1	2	3	4
1	1	0	3	2	5
2	2	3	0	1	6
3	3	2	1	0	7
4	4	5	6	7	0

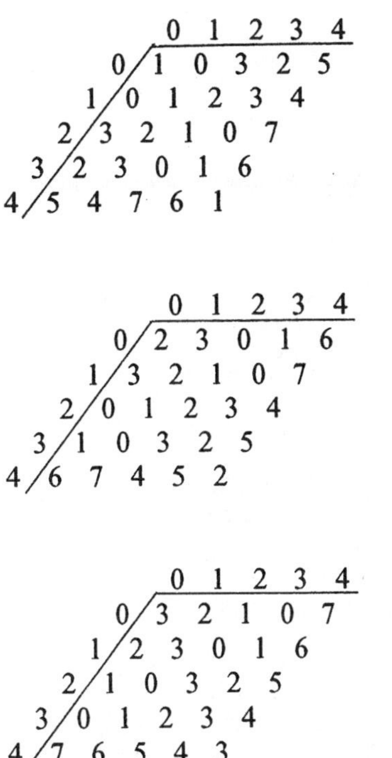

The two-way table can also be useful for three-pile Nim, to find a pile which, with two given ones, produces the Nim-sum 0. This third pile must have a size equal to the Nim-sum of the two given ones. This can be used to find that of the two given piles, whose size ought to be reduced. For instance, start with 3, 5, 7. It is possible to move to

$$3,4,7, \text{ because } 3 \hat{+} 7 \hat{+} 4 = 0$$
$$\text{or to } 3,5,6, \text{ because } 3 \hat{+} 5 \hat{+} 6 = 0,$$
$$\text{or to } 2,5,7, \text{ because } 5 \hat{+} 7 \hat{+} 2 = 0.$$

In this case we have a choice. In other positions we might not have any choice. Starting from 3, 4, 6, we can only move to 2, 4, 6 (Nim-sum 0), but not to 3, 4, 7 (also Nim-sum 0), because a size can not be increased. However, there will always be one pile at least which can usefully be reduced, as we have seen.

We mention, as a possibly useful help for the memory, that the numbers on the same line (straight line or circle) in Fig. II. 4 define positions with Nim-sum 0. They are, in fact, all the winning positions for three piles with sizes not exceeding 7.

(1, 2, 3) (1, 4, 5) (1, 6, 7) (2, 4,6) (2, 5, 7) (3, 4, 7) (3, 5, 6).

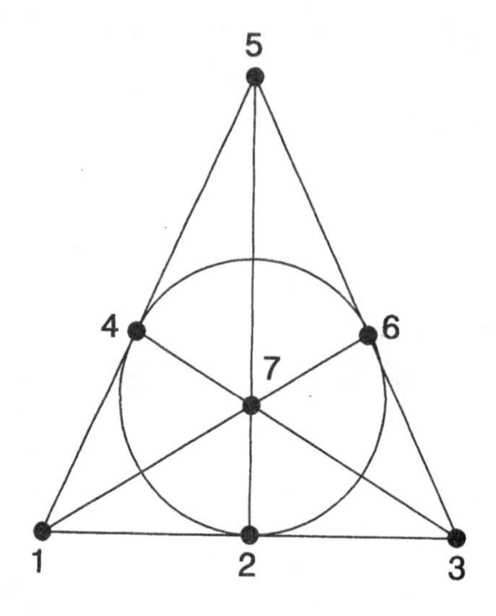

Fig. II.4.

Northcott's Nim

This is played on a rectangular board of squares, with two differet types of counters, belonging respectively to the two players. On each row of the board there are two counters of different types. The players move, in turn, one of their own counters to the left, or to the right, onto an empty square, but without jumping over the other counter. The player who is unable to move any of his counters loses. This can, of course, only happen if each of his counters is on a square at one of the two ends of a row, hemmed in by the opponent's counter on the adjacent square.

Because each player may only move his own counter, this looks like a partizan game. However, in fact it is just another version of Nim. The interval between the two counters in the same row corresponds to the size of a pile, and either player tries to make all these intervals nil by his last move. Admittedly, intervals can be increased as well as decreased, by moving a counter further away from its pair, but in such a case the opponent can restore the length of the interval, and is then nearer to his goal than he was before.

Example

Fig. II.5(a). Intervals 1, 2, 3.

The player whose turn it is will lose, because this happens to be a winning position, which favours the second player. The play will end with position II.5 (b), or with II.5 (c).

The same game can also be played on a strip of squares of finite length, with an even number of counters, no two on the same square. The intervals between the first and the second, the third and the fourth, the fifth and the sixth, and so on, represent the pile sizes.

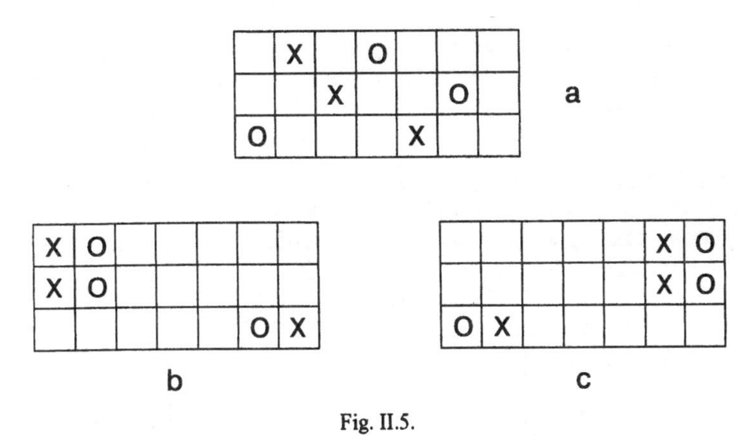

Fig. II.5.

The Silver Dollar Game

This game is due to N. G. de Bruijn. We describe a slightly simplified version, without distorting its essential features. It is played on a semi-finite strip of squares, limited, say, on the left. Single counters are placed on some of the squares. The two players move, alternately, a counter on to an unoccupied square, towards the left, without jumping over another counter. Additionally, another type of move is also provided for: the left-most counter may be removed, and kept by the player who removes it. One of the counters is more valuable than all the others together: this is the 'Silver Dollar'. It is the one which either player will try to remove from the board, to keep it.

If the silver dollar is leftmost to begin with, the first player to move will take it. If there is just one counter to the left of it, at some stage a player will have to take this counter, as the only legal move left for him, and thus he will lose the dollar. But if more than one counter is placed between the edge and the dollar, then imagine one further square to the left of the edge, and play Nim on the strip thus extended.

4.

We can describe Nim with $n > 1$ piles as a game with n components, each component being a single-pile Nim. A move changes only a single component's position. C. A. B. Smith (1968) calls such a game a disjunctive compound of games, while Holladay (1957) calls it a 'Cartesian product'. We shall simply refer to it as to a 'sum' of component games. For other types of compound games consult Smith (1966 and 1968).

In Nim the g-number of a pile is simply its size. In subtraction games and in other one pile games we have computed g-numbers which were not necessarily equal to the pile size, but they had the property that for any smaller g-number a follower position could be found. This followed from the construction of the g-number. We then found that the Nim-sums of the g-numbers of the components had this same propety in n-pile Nim. Any smaller Nim-sum could be constructed by a move affecting just one component.

All this can also be done when the component take-away games are different from one-pile Nim. To find a winning position, we are again interested, in particular, in positions with Nim-sum 0.

Thus the Nim-sum of the g-numbers of the component games may be called the g-number of the sum of the component games. In this sense we may say that any sum of n-pile games with g-numbers is equivalent to a game of Nim (see Appendix V).

As an example, take a sum of S (all Fibonacci numbers) games. As we have seen, the sequence of g-numbers of a component starts with

n	0	1	2	3	4	5	6	7	8	9
$g(n)$	0	1	2	3	0	1	2	3	4	5

Let the initial position consist of piles with sizes 3, 6 and 9. The g-numbers of these piles are (for the subtraction game which we are playing)

	3 ~	11	of pile size 3
	2 ~	10	of pile size 6
	5 ~	101	of pile size 9
Nim–sum		100 .	

To find a winning position, add $100 \hat{+} 101 = 1$ and reduce size 9 to one with g-number 1. This must be possible, because otherwise $g(9)$ could not be larger than 1. So reduce 9 to 1. (8 is a Fibonacci number, viz. F_6.) The result is

3 ~	11
2 ~	10
1 ~	1
	00

An interpolation

It is of interest to note that any sum of *termination games* can be considered to be equivalent to Nim in the sense that a position $(p_1,\ldots,p_n)$ has the 'value' of the position $(i_1,\ldots,i_n)$ in n-pile Nim, if i_j is the 'value' of the component with position p_j).

A termination game is a collection of positions, some of which are terminating, and the rest non-terminating. Two players make moves until a terminating position is reached. The player who produces this position wins, loses, or draws. Also, in any position there is only a finite number of legal moves available.

Clearly, this definition encompasses more games than those we can deal with here. Even Chess is a termination game. To identify a position some features of previous moves would have to be mentioned (whether the king has already moved, for instance), and whose turn it is to move.

The proof of equivalence with Nim, given by Holladay (1957) is an existence proof, and does not tell us how to compute the value of a position. It is therefore not immediately relevant to playing any termination game. In order not to delay further descriptions of games, we quote the proof in Appendix V.

Welter's Game

Counters are placed on a semi-infinite strip of squares, numbered 1, 2, ... with no two counters on the same square. The two players move, turn by turn, one counter to an unoccupied square with a lower number. In the final position, the n counters on the board will be on squares numbered 1, 2,..., n. The player who produces this configuration wins.

If we imagine the numbers on squares occupied by the counters to exceed pile sizes by unity, then we might describe the game as being equivalent to Nim, with an additional condition: no two piles can ever during the play be of equal size.

The phrase 'exceed by unity' had to be added to accommodate the possibility of complete exhaustion of a pile in Nim, and is innocent. But the exclusion of equal pile sizes is essential. We must expect that this condition has an effect on the play, because in ordinary Nim with two piles the winning positions are precisely those with two equal piles.

To begin with, we analyse Welter's game with two counters. The table of g-numbers can be constructed, similarly to the construction of the Nim-addition table, by starting at the top left-hand corner, at configuration (0,1) or (1,0), and computing succesively the smallest non-negative integer not in the set of those integers above or to the left of the number to be entered in the table. There is, of course, no entry in the diagonal, which would refer to two equal pile sizes. We obtain

	$x=$	0	1	2	3	4	5
$y=$	0	.	0	1	2	3	4
	1	0	.	2	1	4	3
	2	1	2	.	0	5	6
	3	2	1	0	.	6	5
	4	3	4	5	6	.	0
	5	4	3	6	5	0	.

The entry for (x,y). is $(x \hat{+} y) - 1$. This is so in the top row and due to the construction of the table, this must remain to be true throughout. Notice that in this formula Nim-addition as well as ordinary addition (or subtraction) are involved.

The winning positions, those with $g\,(x,y) = 0$, are

$$x = 2t + 1,\ y = 2t \quad (t = 0,1,2,\ldots).$$

Now consider the game with $n = 3$ counters. The final position is 1,2,3, which is also a winning position in ordinary Nim, in fact the winning position with the smallest positive pile sizes. Hence we play simply ordinary Nim. There will never be any temptation to make two pile sizes equal, because there is no such winning position in ordinary three-pile Nim, except (0,0,0), which is irrelevant in Welter's game.

Proceed to $n = 4$. This case is mentioned in Sprague (1937) and attributed to Kowalewsky (1930, pp. 55 ff.).

Consider a winning position (a,b,c,d) with $0 \leq a < b < c < d$ in ordinary Nim. We know that it cannot be changed into another winning position in one move, and this is equally true if no two sizes are allowed to be the same.

On the other hand, if (a,b,c,d) is not a winning position, then it is possible to obtain, in one move, a winning one, and it can be done without making two sizes equal, because a position (x,x,y,z), $y<z$, could be changed into the winning poition (x,x,y,y).

It follows from the relationship between ordinary Nim and Welter's game that (a,b,c,d) is winning if and only if $(a-1, b-1, c-1, d-1)$ is winning in ordinary four-pile Nim. For instance, (0,1,2,3) is winning in Nim, and (1,2,3,4) is winning, in fact terminal, in Welter's game.

A solution for $n=5$, but not more than 16 squares on the strip, was suggested by Sprague in 1943 and published in Sprague (1948/9). He attaches to the numbers s_i of the squares labels x_i as follows:

s_i	1	2	3	4	5	6	7	8	9	10	11	12	13	14	15	16
x_i	1	A	B	C	ABC	D	ABD	ACD	BCD	$ABCD$	CD	BD	AD	BC	AC	AB

These labels form a commutative group with four generators of order 2, that is $A^2 = B^2 = C^2 = D^2 = 1$. The label of a position in the game is the product of the five labels of the components. In particular, the label of the terminal winning poition (1,2,3,4,5) is $1 \cdot A \cdot B \cdot C \cdot ABC = 1$.

The labels have the property that if the product of the five labels is not 1, then it is possible by a move, that is by reducing one of the s_i, to obtain a product equal to 1, while if the product equals 1, then no move can reproduce unity. It follows that for any position of the five counters the product of the labels can serve as g-number of the position. The winning positions are those with product 1 of the labels. If the original position has label-product 1, then the opponent can force a win.

Example

$$
\begin{aligned}
&2,3,5,7,8 \cdot A \cdot B \cdot ABC \cdot ABD \cdot ACD = B \\
\to\ &1,2,5,7,8 \cdot 1 \cdot A \cdot ABC \cdot ABD \cdot ACD = 1 \\
\to\ &1,2,3,5,7 \cdot 1 \cdot A \cdot B \cdot ABC \cdot ABD = ABCD \\
\to\ &1,2,3,4,5 \cdot 1 \cdot A \cdot B \cdot C \cdot ABC = 1.
\end{aligned}
$$

The first player wins.

It does not appear that Sprague's ingenious construction of labels can be extended beyond $n=5$, or to more than 16 squares. Welter (1952) has suggested the following sequence of labels

s_i	1	2	3	4	5	6	7	8	9	10	11	12	. .
x_i	1	a	b	ab	c	ac	bc	abc	d	ad	bd	abd	. .

After 2^m steps a new label multiplies the previous labels in sequence. The winning position was again defined by a function of the labels, though not their product, equalling 1. This study referred to any number of squares, but still to not more than 5 counters.

Eventually, Welter (1954) published a complete theory, for any number of counters and any number of squares. The labels are those of the 1952 paper mentioned. The x_i are commutative under multiplication, of order 2.

Let the counters be on squares $s_1, s_2,\ldots s_n$, and let the labels of these squares be $x_{s_1},\ldots, x_{s_n}$. We shall write them $x_1,\ldots, x_n$, without any fear of confusion. Welter defines a function $\varphi(x_1,\ldots,x_n)$ by

(i) $\varphi_n(1, x_1,\ldots, x_{n-1}) = \varphi_{n-1}(Fx_1,\ldots, Fx_{n-1})$

(ii) $\varphi_n(x_1,\ldots, x_n) = x_1^n\,\varphi_n(1, x_1x_2,\ldots, x_1x_n)$

(iii) $\varphi_1(1) = 1.$

Fx_i denotes the label of s_{i-1}.

It follows from (*i*) that the final position $(1,2,\ldots, n)$ has function $\varphi_n(x_1,\ldots, x_n) = 1$ (see the last move in the example below), and from (ii) and (iii) that $\varphi_1(x) = x$.

It is proved that $\varphi_n(x_1,\ldots, x_n) = 1$ is a winning position. The proof uses 12 lemmas and is too long to be given here, but we present a example.

$$\begin{aligned}
2,3,5,7,8 \quad & \varphi_5(a, b, c, bc, abc) = a\,\varphi_5(1, ab, ac, abc, bc) \\
&= a\,\varphi_4(b, c, bc, ac) \qquad = a\,\varphi_4(1, bc, c, abc) \\
&= a\,\varphi_3(ac, ab, bc) \qquad = a.ac.\,\varphi_3(1, bc, ab) \\
&= c\,\varphi_2(ac, b) \qquad = c\,\varphi_2(1, abc) = c\,\varphi_1(bc) = c\cdot bc = b.
\end{aligned}$$

Move to 1, 2, 5, 7, 8.

$$\begin{aligned}
& \varphi_5(1, a, c, bc, abc) = \varphi_4(1, ab, ac, bc) \\
&= \varphi_3(b, c, ac) \qquad = b\,\varphi_3(1, bc, abc) \\
&= b\,\varphi_2(ac, bc) \qquad = b\,\varphi_2(1, ab) = b\,\varphi_1(b) = 1.
\end{aligned}$$

Move to (say) 1, 2, 3, 5, 7.

$$\begin{aligned}
& \varphi_5(1, a, b, c, bc) = \varphi_4(1, a, ab, ac) \\
&= \varphi_3(1, b, c) \qquad = \varphi_2(a, ab) = \varphi_2(1, b) = \varphi_1(a) = a.
\end{aligned}$$

Move to 1, 2, 3, 4, 5.

$$\begin{aligned}
& \varphi_5(1, a, b, ab, c) = \varphi_4(1, a, b, ab) \\
&= \varphi_3(1, a, b) \qquad = \varphi_2(1, a) = \varphi_1(1) = 1.
\end{aligned}$$

It would be desirable, when playing a game, to be aware of a method for finding a winning position which can be reached. But only an iterative trial and error procedure is known.

A different method for solving Welter's game is given in Conway (1976).

k-Welter

This is Welter's game with the further condition that a counter may not move more than k steps down along the strip of squares. (By analogy, $S(1,2,\ldots, k)$ would be called *k*-Nim.)

For a game with two piles we construct again the table of g-numbers as a table of entries equal to the smallest non-negative integer not in the set of up to k entries above or to the left of it. This table depends on k, and we give, as an example, the table referring to $k = 4$.

$s_1 =$ \ $s_2 =$	0	1	2	3	4	5	6	7	8	9	10
0	.	0	1	2	3	4	0	1	2	3	4
1	0	.	2	1	4	0	3	2	1	4	0
2	1	2	.	0	5	1	2	3	0	5	1
3	2	1	0	.	6	2	1	0	3	6	2
4	3	4	5	6	.	3	4	5	6	0	3
5	4	0	1	2	3	.	0	1	2	3	4
6	0	3	2	1	4	0	.	2	1	4	0
7	1	2	3	0	5	1	2	.	0	5	1
8	2	1	0	3	6	2	1	0	.	6	2
9	3	4	5	6	0	3	4	5	6	.	3
10	4	0	1	2	3	4	0	1	2	3	.

The first row is the g-sequence for $S(1,2,3,4)$, shifted one step to the right. Each row is periodic, if we replace the dot by the g-value for $(s_1, s_1 + (k+1))$.

We notice also that the value for (s_1, s_2) equals that for (s_1', s_2') if $s_1 \equiv s_1'$ and $s_2 \equiv s_2'$ (modulo $k+1$); this follows from the periodicity which we have mentioned. Kahane and Fraenkel (1987) have proved that generally, positions $(a_1, \ldots, a_n)$ and $(b_1, \ldots, b_n)$ have the same g-value if $a_i \equiv b_i$ (modulo $k+1$) for $i = 1, 2, \ldots, n$. Their proof is long, and we agree that 'it would be of interest to find a shorter proof'. They show, moreover, that if k is of the form $2^m - 1$, then the g-function equals that of ordinary Welter, reduced modulo $k + 1 = 2^m$. (see also Duvdevani and Fraenkel (1989).)

5.

In the multi-pile games which we have considered the number of piles decreased after a move which removed the last counter of a pile, but the number of piles was never increased. In the games which follow, the number of piles might be increased as the play proceeds (though never the total number of counters).

It appears that at least one of these games has been played for centuries. It is the game called Kayles.

Kayles

According to Dudeney (1907, puzzle 73), the name derived from the French quilles (ninepin) and it was a great favourite in the 14th century.

The play starts with a single 'pile', laid out as a row of counters. The players take turns alternately and remove either one counter, or two adjacent counters from the row, either from one of its ends, or from within. In the latter case the row is split into two, and the game becomes a disjunctive sum of games. Any further move affects just one of the components, reducing it, or splitting it into two. The player who takes the last counter wins.

The g-numbers of positions are computed by the rules for sums of games, as described above, again involving Nim-additions.

Clearly, $g(0) = 0$.

A row consisting of one single counter cannot be split, but the counter can be removed. Hence $g(1) = 1$.

A row of length 2 can be reduced to 1 counter ($g(1) = 1$), or to 0 ($g(0) = 0$). Of course, it cannot be split. Hence ($g(2) = 2$).

A row of 3 counters can be reduced to 1, ($g(1) = 1$), to 2, ($g(2) = 2$), and it can be split, by removing the counter in the middle. This produces ($g(1,1) = 0$). Therefore $g(3)$ equals 3.

We continue and obtain

n	0	1	2	3	4	5	6	7	8	9	10	11	
$g(n)$	0	1	2	3	1	4	3	2	1	4	2	6	...

Guy and Smith (1956) inform us that the sequence is periodic after $n = 71$, with period

412814721827

We illustrate (Fig. II.6) the computation of $g(4)$, which is typical of the computation for any n.

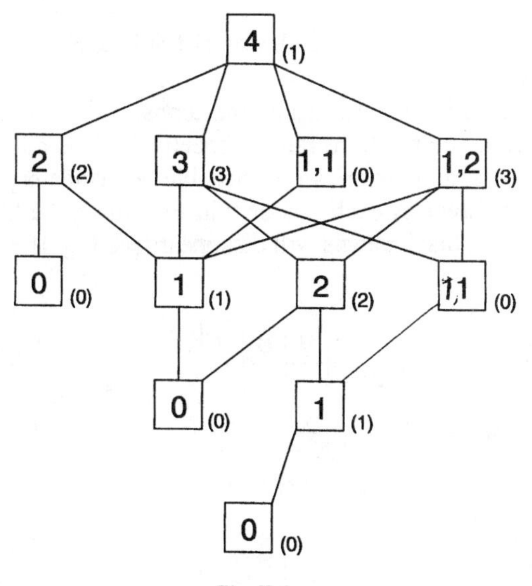

Fig. II.6.

The g-numbers are given in brackets. The winning positions are those with g-number 0. If the play starts with such a position, the second player can win.

It would be awkward to play the game if, for every starting length of the row, a graph had to be constructed. Fortunately, this is unnecessary, because a very simple strategy is available.

If the number in the initial row is odd, the first player removes the central counter, thus producing two rows of equal length. If the initial number is even, he removes the two central counters. Afterwards he 'echoes' the second player's moves, to keep the configuration symmetric. This is analogous to the procedure we use in two-pile Nim.

Not surprisingly, a fair number of games similar to Kayles have been suggested.

Multiple Kayles

If instead of removing one or two adjacent counters, a player may remove n or $n+1$ $(n > 1)$, the sequence of g-numbers is that of Kayles, but each item repeated n times. Hence the name of such games.

If only a single counter were allowed from the edge of the row, without splitting it, then the period of the g-sequence is $\dot{0}\dot{1}$ (as mentioned in Holladay, 1957). A rather more interesting game is Dawson's Chess.

Dawson's Chess

One counter may be removed from an edge of the row, or from strictly inside, together with its immediate neighbours if any. The g-sequence starts with

n	0	1	2	3	4	5	6	7	8
$g(n)$	0	1	1	2	0	3	1	1	0

For $n \geq 52$, the sequence is periodic, with a period of length 34. The name of the game is due to the following scenario:

Consider a chess-board with 3 rows and r columns, with r white and r black pawns, which occupy the first and the third rank, respectively. The pawns move ahead, and take aslant, as in Chess, and taking, when possible, is mandatory. It emerges that if a pawn is moved, one pawn of each colour will eventually be blocked, and the two neighbouring white and black pawns will disappear (see Fig. II.7).

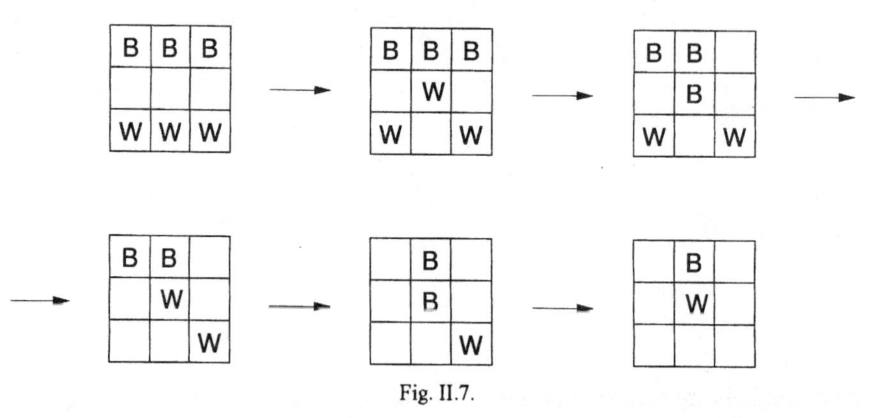

Fig. II.7.

A first version of this game appeared in Dawson (1934), and the version which we describe comes from Guy and Smith (1956). It is, of course, an impartial game, even if the Chess form might suggest otherwise.

The sequence of g-numbers in Dawson's Chess, $g_D(n)$, is closely related to those of other games. The game G_1 (say), where two adjacent counters are being removed, from the edge of the row or from inside, has a sequence starting with

n	0	1	2	3	4	5	6	7	8
$g_1(n)$	0	0	1	1	2	0	3	1	1

that is $g_1(n) = g_D(n-1)$. This sequence is periodic from the 53rd term on, with a period of length 34 (Smith, 1968).

For the game G_2 (say), where only one counter may be removed, strictly from inside the row, the sequence starts with

$$g_2(n) \;\; 0\;0\;0\;1\;1\;2\;0\;3$$

that is

$$g_2(n) = g_1(n-1) = g_D(n-2).$$

The shift of g-numbers is readily explained. The first values are found by direct computation, and the rest is established by induction. Consider

Dawson's Chess	G_1	G_2
with n counters	with $n+1$ counters	with $n+2$ counters.

These row lengths can be changed into 2 rows with length

0 and $n-2$	0 and $n-1$	1 and n
0 and $n-3$	1 and $n-2$	2 and $n-1$
1 and $n-4$	2 and $n-3$	3 and $n-2$
.		
.		
.		

Now take any row of this table. The pairs of g-numbers are the same in the three columns. Hence, if the relationship we want to explain holds up to some n, it will hold for $n+1$ as well, owing to the manner of constructing g-numbers.

Clearly, if n is even in G_1, or when it is odd in G_2, these games can be won by removing first a central set from the row, and then echoing.

Another game with some similarity to Kayles is Twopins.

Twopins (Berlekamp *et al.*, 1982, p. 466)
Counters are set up in columns of 1 or 2, and a legal move takes away either a column of 2, or any two adjacent columns, but never a single column of just one counter.

g-numbers are produced, in the usual way, starting with the terminal positions 0 and 1, which receive $g(0) = g(1) = 0$.

We quote here various positions, with their g-numbers in brackets.

Length of row	positions			
1	1 (0)	2 (1)		
2	1,1 (1);	1,2 (1),	2,1 (1);	2,2 (2)
3	1,1,1 (1);	1,1,2 (2);	1,2,1 (1);	1,2,2 (2);
	2,1,1 (2);	2,1,2 (0);	2,2,1 (2);	2,2,2 (3)

Scrutinizing this list we notice — and we can easily prove it to be generally true — that for computing the g-numbers

—a single 1 at the end of a row can be ignored (1, 2, 2, ~ 2, 2)
—a single 1 between two 2s can be omitted, splitting the row ($2, 1, 2 \sim 2 \hat{+} 2$)
—two 1s at the end of a row next to a 2, or between two 2s are equivalent to a 2.(1, 1, 2 ~ 2, 2; 2, 1, 1, 2 ~ 2, 2, 2).

A graph showing the game starting with 2, 1, 2, 2, might be useful as an illustration (Fig. II.8).

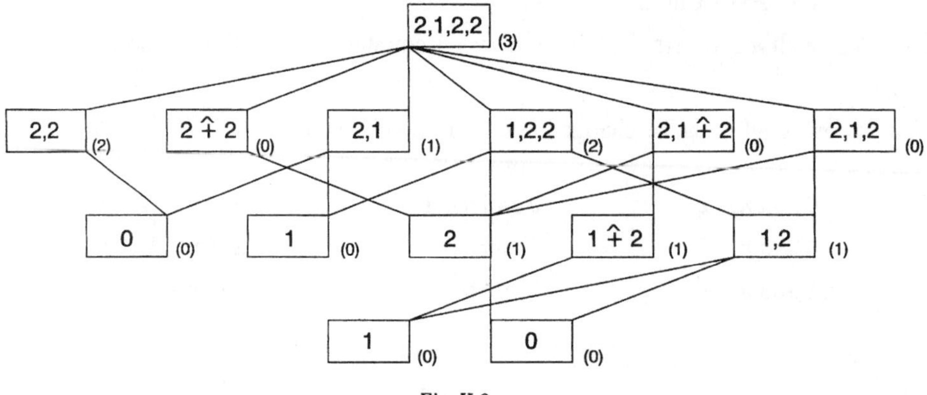

Fig. II.8.

Observe the difference between 2, 2 (one row of two couples) and $2 \hat{+} 2$ (two rows of one couple each).

6.

In the last shown games we split rows of counters by removing one or more counters. In the next group of games we have games where a row may be split without taking away any counters.

The paradigm of such games is Grundy's Game.

Grundy's Game (Grundy, 1939)
We start with one pile of counters. The alternate moves of the two players consist of dividing any pile into two piles of unequal sizes. The player who is unable to move when it is his turn loses. Thus the play ends when there is no pile of more than two counters.

The g-numbers are again computed as for Kayles. Piles (or rows) of size 1 or size 2 have $g(1) = g(2) = 0$. A pile of size 3 can be divided into 1 and 2. Its g-number is $0 \hat{+} 0$, hence $g(3) = 1$. $g(4)$ is the smallest non-negative integer different from the g-number of (1, 3), which is 1, hence $g(4)$ equals 0. So we carry on and obtain

n	0	1	2	3	4	5	6	7	8	9	10	11	12	13	14
$g(n)$	0	0	0	1	0	2	1	0	2	1	0	2	1	3	2

No periodicity has been established for this sequence (to the writer's knowledge).

As a curiosity we mention that B. Descartes (1953) has interpreted the series as a succession of musical notes, and has written words to their music.

A game with moves which divide a pile into equal sizes would not be of interest. It could, of course, only be played with even sizes to begin with, and the first player would win by echoing his opponent's moves as long as posible, leaving eventually the second player with no move to make.

Partitioning
If a move divides a pile into n (>2) portions, then we have to distinguish two cases.

Let the original pile have size s. The two cases are

(a) $s - n$ is even.
(b) $s - n$ is odd.

(a) The first player partitions into $n - 2$ 'piles' of single counters, and the remaining $s - n + 2$ counters into two equal piles. Afterwards he echoes the opponent's moves, and wins. (The piles of size 1 remain, of course, untouched.) If $s - n + 2 < 2n$, he wins with his first move.
(b) The first player divides into $n - 3$ piles of size 1, 1 pile of size 2, and the remaining $s - n + 1$ piles into two equal piles. If $s - n + 1 < 2n$, he wins with his first move.

Lasker's Game
In a collection of board games (Lasker, 1931, pp. 170–230) the mathematician and Chess world champion from 1894 to 1920 quotes the following rules:

A number of rows of counters is given. Any row may be split into two equal or unequal non-zero ones, and alternatively any number of counters may be removed from one row. The player who takes the last counter wins.

This game is obviously of no interest if we start with one single row, but the g-numbers of a single row become relevant, referring to a component of a disjunctive sum of games. The g-sequence for single rows of length n starts with

n	1	2	3	4	5	6	7	8	9	10	11	12
$g(n)$	1	2	4	3	5	6	8	7	9	10	12	11

We might guess that

$$\begin{aligned} g(n) &= n \quad \text{if } n \equiv 1 \text{ or } 2 \text{ (modulo 4)} \\ &= n+1 \text{ if } n \equiv 3 \text{ (modulo 4)} \\ \text{and } &= n-1 \text{ if } n \equiv 0 \text{ (modulo 4).} \end{aligned}$$

In fact, such a guess is correct. To prove it, we have to show that

(a) from every position we can move to one with a smaller g-number or Nim-sum of g-numbers, and
(b) an identical g-number or Nim-sum cannot be reached in one move.

Proof:

(a) From a size with g-number $4t+1, 4t+2$, or $4t+4$, a smaller g-number (or Nim-sum) can be produced by simply reducing the size accordingly. From a size $4t+3$, and g-number $4t+4$ we can obtain a size of a single row with a g-number less than $4t+3$, but not one with precisely $4t+3$, by diminishing the size; however, we can obtain that g-number by splitting into two rows with sizes $4t+2$ and 1 respectively. This can be seen as follows:

The resulting g-numbers will be, in binary notation, ...10 and ....01 respectively. We quote only the last digits, which are unaffected by $4t$, whatever the value of t is. The Nim-sum of the two g-numbers above is of the form .....11, that is congruent to 3 modulo 4.

(b) An identical g-number cannot be produced by reducing a size; therefore we need only investigate the effect of splitting. Since the last two digits are unaffected by t, we prove our contention by arguing with just one value of t, viz. $t = 1$.

Consider, therefore, the sizes 5, 6, 7 and 8, with g-numbers ....01, ...10, ...00, and ....11 respectively.

Now size 5 may be split into (4,1) .. 11 $\hat{+}$ .. 01 = .. 10
or into (3,2) .. 00 $\hat{+}$ .. 10 = .. 10
6 may be split into (5,1) .. 01 $\hat{+}$ .. 01 = .. 00
into (4,2) .. 11 $\hat{+}$ .. 10 = .. 01
or into (3,3) .. 00 $\hat{+}$ .. 00 = .. 00
7 may be split into (6,1) .. 10 $\hat{+}$ .. 01 = .. 11
into (5,2) .. 01 $\hat{+}$ .. 10 = .. 11
or into (4,3) .. 11 $\hat{+}$ .. 00 = .. 11
8 may be split into (7,1) .. 00 $\hat{+}$ .. 01 = .. 01
into (6,2) .. 10 $\hat{+}$ .. 10 = .. 00

$$\text{into } (5,3) \quad ..01 \hat{+} ..00 = ..01$$
$$\text{or into } (4,4) \quad ..11 \hat{+} ..11 = ..00$$

In none of these cases does the original g-number turn into an identical Nim-sum.

Sprague (1935–6) mentions 'without proof' that the g-sequence is the same if splitting into three piles, or into any even number of piles, is also allowed. He states, also, that if the reduction of a row, and splitting into any number of rows is legitimate, then

n	0	1	2	3	4	5	6	$5m+2$	$5m+3$	$5m+4$	$5m+5$	$5m+6$
$g(n)$	0	1	2	4	3	5	6	$8m$	$8m+2$	$8m+1$	$8m+5$	$8m+7$

is the sequence of g-numbers.

If, in another modification of Lasker's game, we may only remove one counter, though also split any row into two, we have

n	0	1	2	3	4	5	6
$g(n)$	0	1	2	0	2	0	2

However, if that one counter must be taken from the inside of the row, thus splitting it, then we find, more interestingly

n	0	1	2	3	4	5	6	7	8
$g(n)$	0	0	1	2	3	1	4	3	2

A comparison with the sequence for Kayles, which we denote now by $k(n)$, discloses that $g(n) = k(n-1)$. This is generally true, as we now show by induction.

Assume that it holds for $n \leq N-1$.

In the preset game, a row of length N may be changed into one of the following positions

$$\begin{array}{ccc} 1, N-1 & & 1, N-2 \\ 2, N-2 & & 2, N-3 \\ \cdot & \text{or} & \cdot \\ \cdot & & \cdot \\ \cdot & & \cdot \\ N-1, 1 & & N-2, 1 \end{array}$$

In Kayles, a row of length $N-1$ may be changed into

$$
\begin{array}{ccc}
0,N-2 & & 0,N-3 \\
1,N-3 & & 1,N-4 \\
\cdot & \text{or} & \cdot \\
\cdot & & \cdot \\
\cdot & & \cdot \\
N-2,0 & & N-3,0
\end{array}
$$

The comparison of corresponding positions in the two games, thus

$$(1,N-1)\ \text{vs.}(0,N-2),\ \ (2,N-2)\ \text{vs.}(1,N-3)\ \ldots$$

and taking into account that $g(n) = k(n-1)$ for $n \leq N-1$, establishes the connection.

Nimania

This game starts with one pile of n counters; in the course of play, more piles might be created, as follows.

Two players, I and II, move alternately by removing one single counter from one of the piles. The player who takes the last counter wins. To begin with, player I reduces the n counters to $n-1$ counters. If n is 1, the play is finished, I has won. But if n is larger than 1, a second pile of $n-1$ counters is created. At the kth move, when some pile has been reduced from m to $m-1$ counters, k more piles of size $m-1$ are attached. The game is finite, because at some stage there will remain a number of piles of size 1, which will gradually disappear. If the number of such piles is odd, the player whose turn it is to move will, eventually, win. If the number of piles is even, the other player will win.

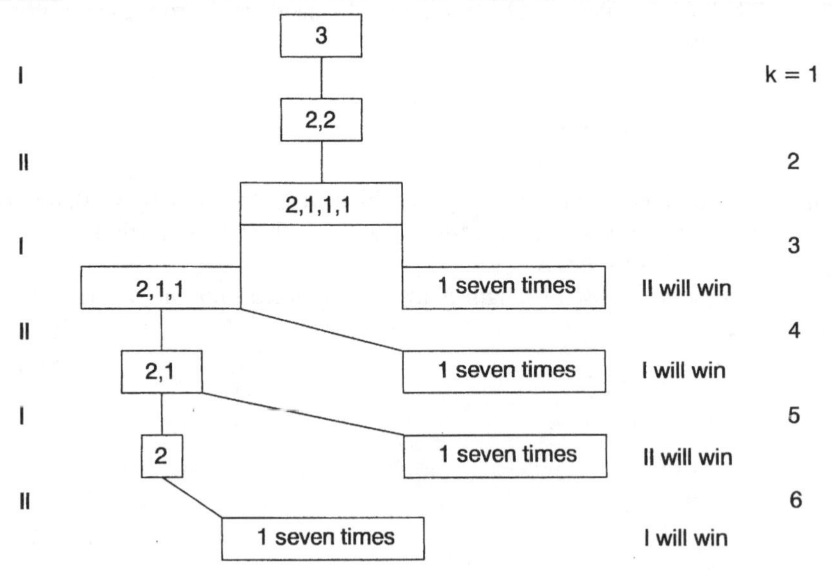

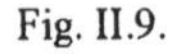
Fig. II.9.

These happenings are meant to simulate, roughly, the spread, increase, and ultimate disappearance of an epidemic.

If $n = 2$, I reduces the pile size to 1 and adds another pile of size 1. II then reduces the position to one pile of size 1, and I takes this single counter and wins.

Fig. II.9 shows that I wins also if the starting pile size is 3.

A. S. Fraenkel and J. Nešetřil (1985), who have invented this game, prove that I has a winning strategy for any n.

Nimania could be described as being played on an initially increasing number of paths. Fraenkel and Nešetřil deal also with this game played on a connected, finite, acyclic graph, when they call the game Coma. For Coma they show that it is finite, and that one of the two players has a winning strategy. Further complications are also envisaged and analysed. A. S. Fraenkel *et al.* (1988) study similar games, when at stage k the number of added copies is $f(k)$, and further results are contained in A. S. Fraenkel and M. Lorberbom (1989).

7.

We have looked at games where moves reduced the size of a pile or a row, and others where a row was split by removing a counter from inside. Other rows were split, even when no counter was removed, but in all these cases a move affected only one single pile or row, a component of a disjunctive sum of games. Now we turn to games where this is not necessarily the case.

Wythoff's game (or Tsyan-Shizi, Chinese Nim) (Wythoff, 1907)
Two players remove alternately a number of counters from one of two piles, or the same number of counters from both piles.

On a two-dimensional plane of co-ordinates we can identify a point (x, y) with a position in which two piles have sizes x and y respectively. On such a plane, we can describe the moves of the players to be Queen's moves on a chess-board, but only to the left, downwards, or diagonally towards the south-west. This is the form in which the game was re-invented by R. P. Isaacs in 1958, as has been pointed out by J. C. Kenyon (1967).

In a two-entry table (x, y) the g-numbers are as follows:

	x	0	1	2	3	4	5
	0	0	1	2	3	4	5
y	1	1	2	0	4	5	3
	2	2	0	1	5	3	4
	3	3	4	5	6	2	0
	4	4	5	3	2	7	6
	5	5	3	4	0	6	8

The player who has secured a position with g-number 0 can carry on with this winning strategy.

It is clearly desirable that we should describe the positions with g-number 0 more directly, without having to construct the table leading up to a number which we want to know. We shall refer to the entries 0 as safe points, and we examine now their properties.

There must be a safe point on every horizontal, on every vertical, and on every diagonal line. Otherwise it would be impossible to move to a safe point from any non-safe one. Also, there can be only one safe point on any such line, because otherwise it would be possible to move from one safe point to another. We can therefore generate the safe points as follows, remembering that a 'diagonal' line refers to the difference of pile sizes of points on it.

The point (0, 0) is safe. It is, in fact the point to be reached in order to win.

The points (1, 2) and (2, 1) are safe.

Increase the difference between the two coordinates by 1, and let the abscissa of the next point be the smallest integer which has not yet appeared as either an abscissa, or an ordinate. Thus we obtain (3, 5), then (4, 7), and so on, together with the symmetrical points (5, 3), (7, 4), and so on.

These safe points can be described, independently of their inductive generation, by their coordinates $([n\tau], [n\tau^2])$, $n = 0, 1, \ldots$, where τ is the golden section $\frac{1}{2}(1+\sqrt{5}) \approx 1.618\ldots$, and $[x]$ denotes the largest integer not exceeding x.

n	$n\tau$	$n\tau^2$	$[n\tau]$	$[n\tau^2]$
1	1.618..	2.618..	1	2
2	3.236..	5.236..	3	5
3	4.854..	7.854..	4	7
	.	.	.	.
	.	.	.	.
	.	.	.	.

To prove this, we must show that

(a) the difference between the coordinates equals n
(b) every non-negative integer appears either as an abscissa or as an ordinate
(c) no abscissa of any point equals the ordinate of any other point.

Proofs.

(a) The difference equals n, because $\tau^2 = 1 + \tau$, and the omitted fractions of $n\tau$ and $n\tau^2$ are therefore equal:

$$[n\tau^2] = [n(1+\tau)] = n + [n\tau].$$

(b) We show that for any n, either $\left[\left[\frac{n}{\tau}\right]\tau\right]$ or $\left[\left[\frac{n}{\tau^2}\right]\tau^2\right]$ equals $n-1$. Assume the contrary, that is $\left[\left[\frac{n}{\tau}\right]\tau\right] < n-1$, and also $\left[\left[\frac{n}{\tau^2}\right]\tau^2\right] < n-1$. (Neither can be greater than or equal to n.) Then

$$\left[\frac{n}{\tau}\right] < \frac{n-1}{\tau}$$

and also

$$\left[\frac{n}{\tau^2}\right] < \frac{n-1}{\tau^2}$$

or

$$\left[\frac{n}{\tau}\right] + \left[\frac{n}{\tau^2}\right] < n-1,$$

because

$$\frac{1}{\tau} + \frac{1}{\tau^2} = 1.$$

But this is impossible, because the sum of two integers is an integer, and the left-hand side differs from $n/\tau + n/\tau^2 = n$ by the sum of two fractions, that is less than 2. So $\left[\frac{n}{\tau}\right] + \left[\frac{n}{\tau^2}\right]$ cannot possibly be $n-2$ or less.

(c) Suppose there could be two integers n and m such that

$$[n\,\tau] = [m\,\tau^2] = s,$$

an integer. Then

$$n\,\tau - 1 < s < n\,\tau$$

and

$$m\,\tau^2 - 1 < s < m\,\tau^2,$$

hence

$$(\,n\,\tau - 1)/\tau + (\,n\,\tau^2 - 1)/\tau^2 < s/\tau + s/\tau^2 < n + m$$

or

$$n + m - 1 < s < n + m.$$

But this is impossible, since s is an integer.

Various modifications of Wythoff's game have been suggested. For instance, Fraenkel and Borosh (1973) have generalized it as follows: instead of allowing any number of counters to be removed from a single pile, only a multiple of b may be removed. Alternatively, remove $k > 0$ and $l > 0$, respectively, from the two piles, subject to $k - l$ being a multiple of b, and $|\,k - l\,| < a \cdot b$, where a and b are given parameters.

These rules have the effect that if the difference between the two pile sizes is congruent to r modulo b to begin with, then this congruence prevails throughout a play.

Of course, $a = b = 1$ would be the original Wythoff's game.
The player who cannot make a legitimate move loses. This happens, for instance, if the position consists of only one pile with size less than b.
Fraenkel and Borosh have proved that winning positions are those where the two pile sizes $\Phi_n(r)$ and $\Psi_n(r)$ are

$$\Phi_n(r) = [n\alpha + \gamma(r)], \quad \Psi_n(r) = [n\beta + \delta(r)]$$

$r = 0,1,2,\ldots$ and

$$\alpha = \tfrac{1}{2}(2 - ab + \sqrt{a^2 b^2 + 4}), \quad \beta = \tfrac{1}{2}(2 + ab + \sqrt{a^2 b^2 + 4}), \quad \gamma(0) = \delta(0) = 0$$

$$\gamma(r) = \frac{1}{ab}(b\,(a-1) + \alpha r) \quad (0 < r < b)$$

$$\delta(r) = \frac{1}{ab}(b\,(a-1) + \beta r) \quad (0 < r < b).$$

Let us now look, in particular, at games where $a = 1$ (this case was introduced by Connell, 1959) and $b = 2$.

Now $\alpha = \sqrt{2}$, $\beta = 2 + \sqrt{2}$, $\gamma(1) = 1/\sqrt{2}$, $\delta(1) = 1 + 1/\sqrt{2}$

$$\Phi_n(0) = [2n/\sqrt{2}], \quad \Psi_n(0) = [2n(1 + 1/\sqrt{2})]$$

$$\Phi_n(1) = [(2n+1)/\sqrt{2}], \quad \Psi_n(1) = [(2n+1)(1 + 1/\sqrt{2})]$$

The tables in which we find the winning positions, those with entries 0, are, for pile sizes m and n

r = 0 (even difference)

n \ m =	0	1	2	3	4	5	6	7
$n =$ 0	0	.	1	.	2	.	3	.
1	.	1	.	0	.	3	.	2
2	1	.	2	.	3	.	0	.
3	.	0	.	3	.	2	.	1
4	2	.	3	.	4	.	5	.
5	.	3	.	2	.	5	.	4
6	3	.	0	.	5	.	6	.
7	.	2	.	1	.	4	.	7

r = 1 (odd difference)

n \ m =	0	1	2	3	4	5	6	7
$n =$ 0	.	0	.	1	.	2	.	3
1	0	.	1	.	2	.	3	.
2	.	1	.	2	.	0	.	4
3	1	.	2	.	3	.	4	.
4	.	2	.	3	.	4	.	5
5	2	.	0	.	4	.	5	.
6	.	3	.	4	.	5	.	6
7	3	.	4	.	5	.	6	.

If the rules are changed, so that either precisely two counters may be taken from a single pile, or precisely one counter from both piles, then in these tables any even number is to be replaced by 0, and any odd number by 1.

These are the first seven winning positions

$2n$	$2n\alpha$	$2n\beta$	$\Phi_n(0)$	$\Psi_n(0)$	$2n+1$	$(2n+1)\alpha$	$(2n+1)\beta$	$\Phi_n(1)$	$\Psi_n(1)$
0	0.0000	0.0000	0	0	1	0.7071	1.7071	0	1
2	1.4142	3.4142	1	3	3	2.1213	5.1213	2	5
4	2.8284	6.8284	2	6	5	3.5355	8.5355	3	8
6	4.2426	10.2426	4	10	7	4.9497	11.9497	4	11
8	5.6568	13.6568	5	13	9	6.3639	15.3639	6	15
10	7.0710	17.0710	7	17	11	7.7781	18.7781	7	18
12	8.4841	20.4841	8	20	13	9.1923	22.1923	9	22

Knights

We have mentioned that moves in Wythoff's game can be said to be Queen's moves in Chess, restricted to certain directions. Similarly, we might think of knight's moves, where the moves allowed are as in Fig. II.10. This is the same as a game with two piles of counters, where the legal moves are: either remove two counters from one pile and one from the other, or remove two counters from one pile and transfer one of them to the other pile.

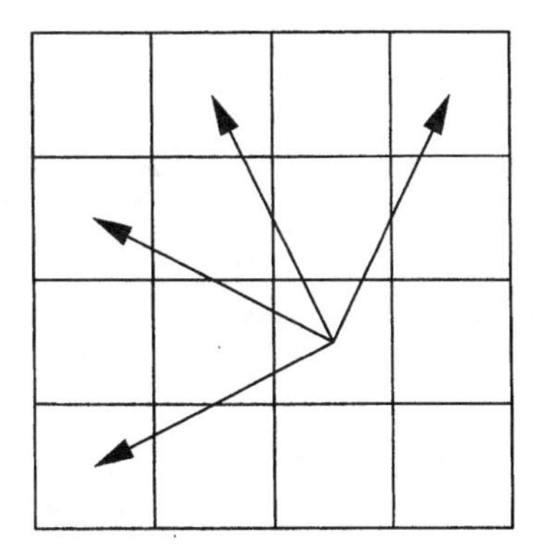

Fig. II.10.

The following table contains the g-numbers:

	1	2	3	4	5	6	7	8	9
1	0	0	1	1	0	0	1	1	0
2	0	0	2	1	0	0	1	1	0
3	1	2	2	2	3	2	2	2	3
4	1	1	2	1	4	3	2	3	3
5	0	0	3	4	0	0	1	1	0
6	0	0	2	3	0	0	2	1	0
7	1	1	2	2	1	2	2	2	3

The winning positions, those with g-number 0 are

$$(4n+1, 4m+1),\ (4n+1, 4m+2),\ (4n+2, 4m+1),\ (4n+2, 4m+2),$$
$$(n,m = 0, 1, 2, ..)$$

Moore's game

E. H. Moore (1910) has proposed the modification of Nim, where a player may remove counters from up to k piles at the same time. The player who takes the last of all counters wins. (Of course, we assume that the number of piles initially exceeds k.) We call this game Nim(k). Ordinary Nim is then Nim(1).

Moore has given the following winning strategy: express the pile sizes in binary notation, as in Nim(1), add the digits of the same power of 2, but this time reduce the sum modulo $k+1$, rather than modulo 2, as in Nim (1). The winning positions are again those in which all digits of the reduced sum are zero.

Moore's solution has been quoted in many later descriptions of the game, but the present writer has not seen any rigorous proof; he might possibly have overlooked one. In any case, we present a proof here.

As was the case for Nim(1), we must show that

(a) a move can change a winning position only into a non-winning one.
(b) a non-winning position can be changed into a winning position.

(a) This is obvious, just as for Nim(1).
(b) Consider the column of the highest power of 2 in which there appears the digit 1 for a pile size (which was expressed in binary notation). In the sum, any of the numbers 0, 1,...k might appear.

If that number is 0, proceed to the next column, for a lower power of 2.

To begin with, assume that the number in the sum is k. This will be so if the digit 1 appears $n(k+1)+k$ times in the pile sizes ($n = 0, 1,...$). Choose k of those pile sizes (rows) in which the digit is 1, and call them the 'affected' rows. As part of the move, change 1 in these rows, and in the column dealt with, into 0. The corresponding digit in the sum is thereby reduced to 0, but only in this column.

For instance, let the pile sizes be 7, 6, 6, 3, 2, and let k equal 3, $k+1 \equiv 4$.

$$
\begin{array}{rcrcr}
7 & & 111 & & 11 \\
6 & & 110 & & 10 \\
6 & \approx & 110 & \rightarrow & 10 \\
3 & & 11 & & 11 \\
2 & & \underline{10} & & \underline{10} \\
 & & 312 & & 12
\end{array}
$$

This is the first step in the move. We must yet deal with the other columns, unless all their sums are already 0.

Observe that if in an affected row we change, in the remaining columns, a 1 into a 0, or a 0 into a 1, we shall still have reduced the size, after we have performed the first step just described. In the other rows, changing a 0 into a 1 would mean an illegal increase.

If in one of the columns yet to be dealt with the sum s is positive, then let there be in the affected rows m 1s and $n = k - m$ 0s. Apply the following rules:

— If $m \geq s$, remove s 1s. The digit in the sum becomes 0.

— If $m < s$, add 1 to $k - s + 1$ of the n 0s. Again, the digit in the total will be 0.

We continue with our example.

$$
\begin{array}{l}
\text{affected rows} \\
\text{after the} \\
\text{first step}
\end{array}
\quad
\begin{array}{lcccccl}
\left\{\begin{array}{l} 11 \\ 10 \\ 10 \end{array}\right. & & \begin{array}{c} 01 \\ 10 \\ 10 \end{array} & & \begin{array}{c} 01 \\ 11 \\ 11 \end{array} & & \begin{array}{l} =1 \\ =3 \\ =3 \end{array} \\
\;\;11 & \rightarrow & 11 & \rightarrow & 11 & & =3 \\
\;\;\underline{10} & & \underline{10} & & \underline{10} & & =2 \\
\;\;12 & & 02 & & 00 & & \\
m > s & & m < s & & & &
\end{array}
$$

We have thus dealt with the case when the left most digit in the sum is k.

Now let that digit be $t < k$. Then remove, in the first step of the move, only t of the 1s, in affected rows. If in any of the other columns the sum is $s > 0$, and there are m 1s and n zeroes, $m + n = t$ in the affected rows of those columns, apply the following rules:

— If $m \geq s$, remove s of the 1s, as before.

— If $m < s$, distinguish between (i) $k - t \geq s - m$, and (ii) $k - t < s - m$. If (i), remove the m 1s in the affected rows, and another $s - m$ in some other rows. If (ii), add 1 to $k - s + 1$ of the n zeroes in the affected rows.

Example

$$
\text{affected rows}
\quad
\begin{array}{rcccccl}
\left\{\begin{array}{r} 7 \approx 111 \\ 4 \approx 100 \end{array}\right. & & \begin{array}{c} 11 \\ 00 \end{array} & & \begin{array}{c} 11 \\ 10 \end{array} & & \begin{array}{l} 10 \approx 2 \\ 10 \approx 2 \end{array} \\
3 \approx \;\;11 & & 11 & & 11 & & 10 \approx 2 \\
2 \approx \;\;\underline{10} & \rightarrow & \underline{10} & \rightarrow & \underline{10} & \rightarrow & \underline{10} \approx 2 \\
232 & & 32 & & 02 & & 00 \\
 & & \text{(ii)} & & \text{(i)} & &
\end{array}
$$

This proof is an existence proof; we do not suggest that in an actual play an experienced player will use these very moves.

Now we shall consider the case of a disjunctive sum of Moore's games, and the problem of finding Grundy numbers for Nim(k), such that their Nim-sum is the g-number of the sum of such games. As Jenkyns and Mayberry (1980, p. 51) remark, such a function is extremely complicated to compute for more than k piles. We quote their result for the simplest of these cases, that is for Nim(k) with $k + 1$ piles.

Let the sizes of the $k + 1$ piles be $p_0\, p_1\, \ldots\, p_k$. We call p_0 the base, and

$$e = \sum_{i=1}^{k} (p_i - p_0)$$

the excess of this position. If the binary coefficient $\binom{e+1}{2}$ is smaller than the base, the g-number equals

$$\binom{e+1}{2} + \left(p_0 - \binom{e+1}{2} - 1\right) (\text{modulo } e + 1)$$

If $\binom{e+1}{2} = p_0$, the Grundy number equals $\sum_{i=0}^{k} p_i$.

Their paper includes also an example of a disjunctive sum of $N(3)$ games, each with 4 piles.

Rosebushes

This is Moore's game with the further condition that only one single counter may be removed from any of the k or fewer affected piles.

Let m be the largest integer not exceeding $\frac{1}{2}k$. Then the winning positions are (see Schuh, 1968) (writing the n pile sizes in not decreasing order, and writing e for even and d for odd sizes):

when $n = m + k + 1$

$$\underbrace{\text{ee..e}}_{n \text{ times.}} \quad \text{and} \quad \underbrace{\text{dd..d}}_{k+1}\underbrace{\text{ee..e}}_{m \text{ times.}}$$

and when $n = m + k + 2$

$$\underbrace{\text{ee..e}}_{n \text{ times.}} \qquad \underbrace{\text{dd..d}}_{k+1}\underbrace{\text{ee..e}}_{m+1 \text{ times.}}$$

$$\underbrace{\text{ee..e}}_{m+1}\underbrace{\text{dd..d}}_{k+1} \qquad \underbrace{\text{dd..d}}_{k-m}\underbrace{\text{ee..e}}_{m+1}\underbrace{\text{dd..dd}}_{m+1 \text{ times.}}$$

In particular, when $k = 1$, hence $m = 0$, this means

when $n=2$ ee and dd

when $n=3$ eee dde ded edd.

This is equivalent to the disjunctive sum of $S(1)$ components, where, as we have seen, the g-number is 0 for even and 1 for odd sizes.

For the cases $n = m + k + t$, $t > 2$ the general solution is (as yet) unknown, except of course for $k = 1$, $n = t + 1$.

This game appears, with $k = 2$, in a tale about Princess Romantica, whose roses are picked by two princes. The prince who wins the game wins the hand of the princess (de Carteblanche, 1970, 1974).

Examples

$n=5$ $k=2$ $m=1$ $t=2$

```
1 1 2 3 4
1 1 1 2 4*
1 1 1 2 3
1 1 1 2 2*
1 1 1 1 1
0 0 1 1 1*
0 0 0 1 1
0 0 0 0 0*
```

$n=5$ $k=3$ $m=1$ $t=1$

```
1 1 2 3 4
1 1 1 3 4*
1 1 1 2 4
0 0 0 2 4*
0 0 0 1 4
0 0 0 0 4*
0 0 0 0 3
0 0 0 0 2*
0 0 0 0 1
0 0 0 0 0*
```

*winning positions.

The most complex rules for moves affecting more than one pile or row known to this author are to be found in N. Y. Wilson (1959). It concerns a disjunctive sum of components where each player may remove

(1) one counter from any component (possibly splitting it) or
(2) three counters from anywhere (possibly splitting components) or
(3) two adjacent counters from any component.

N. Y. Wilson does not say where the inspiration for just these rules came from, but he asserts that there are only 14 different values in the sequence of g-values (as we would say), and that for $n = 53$ the sequence is 'periodic' with a length 34 of the period in a sense explained in the paper. The reader might notice that this is also true of Dawson's Chess, but the connection between Dawson's game and Wilson's is not elucidated.

8. ENTAILING

This concept is introduced in Berlekamp *et al.* (1982, p. 376).

An entailing move forces the opponent to move on the same pile on which the last move was done. This has various consequences for playing correctly. Suppose that removing 1 counter entails, whichever player has moved. Then the player who reduces any one of the piles to one counter, will lose. The opponent will take this counter, entailing, but will thereby win, because there is nothing left in the pile from which the opponent ought to take at least one. Any other piles, which might have been left, are irrelevant.

In the modified Lasker's game, where one counter may be removed from a pile, or any pile might be split into two, the player who leaves three counters in a pile, loses, again assuming that removing one counter entails. The expert opponent will reduce the pile size to two, entailing. He leaves to the first player the choice of reducing the pile size to 1, thus losing as explained above, or to split the pile into two piles of 1 counter each. Then he loses again, because the opponent will entail one of these piles by taking one.

9.

Take-away games are of the most varied types and forms. Here is one due to Gale (1974), which he called Gnim.

Gnim

$n \cdot m$ counters are laid out in a rectangle with n rows and m columns. The two players remove alternately a counter together with all those above and to the right of it. The player for whom only one counter is left to remove loses.

We have here, for the first time in this book, a game for which we can prove that the player who moves first has a winning strategy for any n and m, though we cannot say what this strategy is, except in special cases.

Assume that the second player has a winning strategy. We show that this is impossible.

Let the first player A start by taking away the top rightmost corner. B answers it with a winning move — we have assumed that he has one. This produces a position which A could have produced by his first move, thus starting a winning sequence of moves. But it is imposible that A as well as B should have a winning strategy, so we must reject our original assumption that B has got one. One of the players must have a winning strategy, and this player must be A.

This argument does not tell us anything about the winning move or strategy. We can indicate it, though, for special cases.

Case I: n = 2

We start by removing the top rightmost corner. This leaves the lower row with just one more counter than the top one. Such a position can be produced by A, whatever B does. Eventually one single counter remains, in the lower row, and B must take it, and lose.

Of course, the role of row and column can be reversed.

Case II: m = n
A starts by leaving only the leftmost column and the bottom row. After this, whenever B takes c counters from the row (or column), A takes c counters from the column (or row).

Gardner (1973) mentions unique first winning moves for the following cases.

Case III: Four times five board

```
0 0
0 0
0 0 0 0 0
0 0 0 0 0
```

Case IV: Four times six board

```
0 0
0 0
0 0
0 0 0 0 0 0
```

In other cases the first player may have more than one winning opening. The smallest rectangle with more than one winning strategy for the first player has been found, by a computer programme, to be that with 8 rows (or columns) and 19 columns (or rows).

When we discussed Whythoff's game we mentioned that it was re-invented, in an equivalent form, by R. Isaacs. A similar incidence of an independent discovery of the same game can be related with regard to Gnim. This was noticed by an observant reader of M. Gardner's column in the *Scientific American*. About a quarter of a century before Gale's invention, Schuh (1952) proposed his Game of Divisions.

Game of Divisions
A positive integer N is given. List all divisors of N, including 1 and N itself. Turn by turn two players delete a number in the list, together with all its divisors. The player who must delete N loses.

If N has only two prime factors, then this is clearly the same as Gnim. For instance, take $N = 675 = 3^3 \cdot 5^2$. We write the list of divisors as follows

	3^3	3^2	3^1	3^0
5^0	27	9	3	1
5^1	135	45	15	5
5^2	675	225	75	25

The proof that the first player can win was also known to Schuh. His version has the advantage that it is simple to envisage the game in higher dimensions, when N has more than two prime factors. For cubical Gnim it turns out that on a n^3 cube the winning opening consists of taking a $(n - 1)$-cube from the appropriate corner when n equals 2 or 3.

If, together with a divisor of N, all the multiples of that divisor, not exceeding N, are to be removed, and the player removing 1 loses, then the game is essentially Gnim — turned upside down, as it were. Such an alteration does not produce anything new. However, we can devise a new game.

New Game

Given N, any number not exceeding N may be removed, with all its divisors. The player who removes N loses, as before. The proof that the first player has a winning strategy carries over from that for Gnim.

With sufficient time, and a sufficiently large sheet of paper, the winning strategy can be found for any N. For small N we can find it by inspection, but no general pattern emerges.

During such inspection it is useful to realize that if a player is faced with N and an even number of other remaining integers, none of them having still left a divisor, then that player loses. For instance, if N equals 8, and the remaining numbers are 4, 5, 6, 7 and 8, but no divisor of 4 or of 6, then the players will alternately take one of 4, 5, 6 and 7, and the next player will take 8 and lose.

The following is a list of winning moves for small values of N

N	2	3	4	5	6	7
A takes	1	2,1	1	2,1	4,2,1	6,3,2,1
leaving	2	3	2,3,4	3,4,5	3,5,6	4,5,7

so that B must lose. For instance, take the case when N equals 4. A takes 1, and leaves 2, 3, and 4. If B takes 2 (or 3), A takes 3 (or 2) and B must take the remaining 4. On the other hand, if B took 2, 4, he would thereby lose immediately.

When N equals 8, A should take 1; the play will continue in one of the following ways:

A has left 2, 3, 4, 5, 6, 7, 8.

If B takes	2	3	4,2	5	6,3,2	7
then A takes	3	2	6,3	6,3,2	5	6,3,2
leaving	4,5,6,7,8	4,5,6,7,8	5,7,8	4,7,8	4,7,8	4,5,8

In all these cases B will lose.

Observe that it follows that if A had started with any other move than taking 1, he would have lost, as the table shows. (1 is taken away at the first move in any case.)

In this game we are not bound to remove only divisors of N, as was the case in Gnim. We can also omit the condition that the player who removes N loses, and stipulate instead that a player must remove, with any number, all its divisors, and that he who removes the last remaining number, whichever this is, loses.

Observe that this modification does not alter anything if N is a prime, because since it has no divisors, it can be removed at any stage, or might be just left last.

If we scrutinize now the positions for small N we find the following optimal first moves for the first player

N	2	3	4	5	6	7
A takes	1	2,1	4,2,1	2,1	1	6,3,2,1
leaving	2	3	3	3,4,5	2,3,4,5,6	4,5,7

The case $N=6$ is not as obvious as the other cases, so we exhibit the complete argument for the first player, who starts with 1, and leaves 2, 3, 4, 5, 6.

B takes	2	3	4,2	5	6,3,2
A takes	3	2	6,3	6,3,2	4
leaving	4,5,6	4,5,6	5	4	5

In all these cases A must eventually win.

In Gnim it would not make any sense to make the removal of N a winning move. A would simply take the lot. But this is not so in the game where, given 1,2,..., N, the players remove turn by turn any number not exceeding N with all divisors of the number taken, and the player who takes the last number *wins*. As before, the first player has a winning strategy, and the proof follows that for Gnim. For small N, these are now A's optimal first moves

N	2	3	4	5	6
A takes	2,1	1	2,1	4,2,1	6,3,2,1
leaving	—	2,3	3,4	3,5	4,5

A will win.

Sylver Coinage

Two players in turn call a positive integer not expressible as the weighted sum of any of the previous integers called, nor must it be 1. The example below shows that this progressively restricts the players' choice until finally only 1 is left. In that case the player whose turn it is loses.

Example

Let the initial choices of players A and B be $a=4$, $b=7$, respectively. If A begins, his choice lies among the integers n such that no integer multipliers can be found to give $n=4f+7g$. It is easily confirmed that such integer multipliers constitute the set S = (1, 2, 3, 5, 6, 9, 10, 13, 17). Suppose A then chooses 5. This excludes from S the numbers 8, $10=2\times 5$, $13=2\times 4+5$, $17=2\times 5+7$, and leaves the reduced set $S'=(1,2,3,6)$ for B to choose from. Now we see that if B chooses 2, and also excludes thereby 6, A chooses 3 and wins, while if B chooses 3 and excludes thereby also 6, A wins by choosing 2.

The winning strategy derives from the following theorem of Sylvester (1884). The game is due to J. H. Conway, but the spelling Sylver credits Sylvester.

Theorem
If a and b are relatively prime, the set of those numbers that cannot be expressed as a sum of multiples of a and b is finite. The largest of this set is $a \times b - a - b$ (which equals 1 if a and b are 2 and 3). Each time a valid number is called, the eligible set is reduced and will eventually contain only 1. Therefore the game cannot continue forever.

If the first player calls 2, the second player wins by calling 3 (he has to), and vice versa. R. C. Hutchings has proved that by choosing a number larger than 3, the first player can win. His proof is not constructive. Guy (1976) has given further information about the game, and lists a number of questions about it.

10. MISÈRE GAMES

In most take-away games we have been dealing with it was the player making the last move who won. We turn now to the so-called misère form of take-away games: the player who makes the last move loses. In fact, it is this form of Nim which is most popular in many countries.

As Grundy and Smith (1956) have stressed, the theory of Misère games is rather complicated, and rudimentary. We shall merely give some impression of the way of manipulating such situations.

In one-pile Misère-Nim the situation is, of course, as trivial as in ordinary Nim. The first player takes all but one of the counters, if n is larger than 1. If n is 1, he loses by having to take it. We start by describing winning positions of disjunctive sums of subtraction games $S(s_1,...)$.

Let the pile sizes be $x_1, x_2, ..., x_n$. Such a position is winning if (denoting the g-numbers by $g(x_1),...$)

$$\text{either } g(x_1) \hat{+} g(x_2) \hat{+} \dots \hat{+} g(x_n) = 0, \text{ and at least one } g(x_i) \geq 2. \qquad (P_1)$$

$$\text{or if } g(x_1) \hat{+} g(x_2) \hat{+} \dots \hat{+} g(x_n) = 1, \text{ and all } g(x_i) \text{ are either 0 or 1.} (P_2).$$

Note that in ordinary Nim the position (P_1) is winning, and that a position (P_2) consists of an odd number of piles, each with one single counter. Also, a position (P_1) will have more than one pile with g-number at least as large as 2; otherwise the Nim-sum could not be 0.

To prove our statement about winning positions we show, first, that

(i) all followers of (P_1) and of (P_2) are non-winning and that
(ii) all non-winning positions have at least one follower which is either a position (P_1), or a position (P_2) (Ferguson, 1974).

Proofs.
(i) Any follower of a (P_1) position will have a Nim-sum different from 0. It follows from our remark above that such a follower will have at least one pile with g-value not smaller than 2. It will therefore be neither in (P_1), nor in (P_2).

Now consider followers of a position (P_2). If a follower has still only components with g-values 0 or 1, then its Nim-sum must have changed from 1 into 0, and it is not in (P_1), nor is it in (P_2). But if it has a component with g-value ≥ 2, and is thus not in (P_2), then it cannot have Nim-sum 0, because the other components have retained their values: the follower is not in (P_1) either. This proves (i).

(ii) If a non-winning position has some component with g-value not smaller than 2, then there is a follower with g-value 0, as we know from the theory of the ordinary (last move wins) case. This follower is in (P_1), if some component has g-value at least 2, by definition of (P_1). Otherwise, that is if all components of the follower have value 0 or 1, ten the original positions could have only a single component, C say, with value not smaller than 2, all other components having value 0 or 1, and the follower can be made into having total Nim-sum 1, thereby being in (P_2).

If, however, a non-winning position has all its components with value 0 or 1 and an even number with value 1 (otherwise it would be winning), then by Ferguson's theorem about subtraction games a follower in (P_2) can be found. This proves (ii).

It will be appreciated that for Misère-Nim the proof could have been simpler. It is already, in essence, contained in Bouton (1902). But we have chosen our approach so as to make the statement valid for all take-away games. For Misère-Nim our result means: play as for ordinary Nim, until you can move to a position in which all piles have just one counter. Then move to a position which has an odd number of such piles.

We have seen that Ferguson's theorem was crucial in our proof. Ferguson (1974) discusses also necessary and sufficient conditions for positions to be winning in other than subtraction games.

B. GAMES PLAYED ON GRAPHS OR BOARDS

1. TREES AND WOODS (see Appendix iv.)

Hackendot

This game was invented by John von Neumann and discussed by Úlehla (1980). A set of directed trees (a directed forest) is given. Each of the trees has its own root. A move consists of removing a vertex of one of the trees, together with all edges and vertices on the path from the root to that vertex, together with the root itself and all adjoining branches. Each move establishes a new forest. The two players move alternately, and the player who takes the last vertex wins.

We shall present a winning strategy, but before we do this we prove, following von Neumann, that if initially we have only one single tree, then the first player has a winning strategy.

Suppose, to the contrary, that it is the second player who has the winning strategy. Let A, the first player, select for removal the root alone. The second player, B, answers according to his winning strategy, generating a winning position. But A could have produced that position by his first move. Since it is impossible that both players should have a winning strategy, we must abandon the assumption that it is B who has got one. It must be A who can force a win.

This argument does not tell us which that optimal strategy is. In what follows, such a strategy is presented, for any forest.

Consider the forest on which the game is being played. Now imagine that this forest describes the disjunctive sum of some abstract games. This is, of course, quite different from Hackendot. But the idea serves to attach to the nodes labels, which would be the g-numbers of the nodes which are positions in this abstract game. In fact, all that matters in the present context is whether such a number is zero or positive. We call the nodes with zero label white, the others black. All terminal nodes are white.

Our concern is the construction of Grundy numbers (Nim-sums) of the forest positions of Hackendot. For this purpose, construct for a forest F another forest LF as follows: it contains all black nodes of F, with the following structure.

b is a follower of a, if it was a follower of a in F, or the follower of a white node in F which was itself a follower of a (see Fig. II.11). The sequence $L(LF) = L^2F, L^3F,...$ must terminate, because at each step the number of nodes is being reduced. Denote F by L^0F.

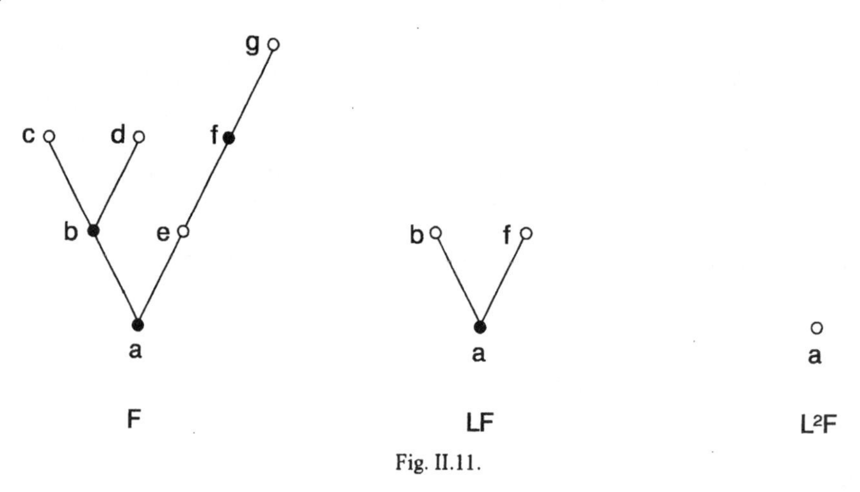

Fig. II.11.

Now find in each forest L^kF ($k = n, n-1,...,0$) the number of white roots. If this number is even, write $R_k = 0$, otherwise write $R_k = 1$. The g-number of the original forest F is, in binary notation,

$$R_nR_{n-1}\ldots R_1R_0.$$

Úhlela's proof is simple, but too long to be reproduced here. An example will be welcome — see Fig. II.11. The number of white roots is 0, 0, 1, therefore the g-number of the forest (in this example actually a single tree) is 100 ~ 4.

Knowing how to construct the g-numbers is of little use during a play, where we are looking for a node to remove, so as to reach a winning position. Úhlela (1980) shows how to do this in the course of his proof for the construction of g-numbers. We sketch here his procedure.

Given F, find in the sequence L^kF $(k = 1,2,...)$ the largest k for which $R_k = 1$. Remove one of the white roots, say N_k. Look at $L^{k-1}F$ and decide whether to remove N_k or one of its white followers, to obtain $R_{k-1} = 0$. Say you choose N_{k-1}. In this manner work back to L^0F and remove N_0. This produces a winning position.

2.

From trees and woods, we turn to other geometrical forms: for instance to circles. First, a simple example.

Round Table

Girls and boys should sit on chairs round a table. B leads boys, and G leads girls to a seat. No boy must sit next to a girl. B and G play this as a game. They move alternately, B first, and he who cannot find a seat for his charge loses.

The winning strategy depends on the number of chairs. Let B start the play. If the number of chairs is even, G will win by seating a girl opposite to the boy who has just taken his seat. G will have the last move. (Some chairs will have to remain empty.)

If the number of chairs is odd, then there is no opposite chair. In this case B can win by choosing first any chair, and afterwards he acts as if this chair did not exist, and as if he were the second player. (If there are only three chairs, only one boy can be seated and sits alone at the table.)

Curiously, up to five places, whoever has the winning strategy cannot lose by making a mistake. He cannot possibly make any move except the recommended one.

If, instead of seats around a circle, we have a seat in the centre as well, and we call the centre a neighbour of all the other places, then the first player wins by starting in the centre. If to avoid this triviality, we stipulate that the first move must not use the centre, then we are back at the game without a centre, because the latter will never be able to be used, even at any later stage.

Mixed circle

Now consider the rule opposite to that for Round Table: no children of the same sex must sit next to each other. Then the second player can always win by putting his charge next to the child which has just been seated, either consistently clockwise, or consistently anti-clockwise.

Marguerite

Number the vertices of a polygon clockwise $1,2,...n$. The first player puts a counter on vertex 1. From then on the players count, alternately, m_1, or $m_2,...,$ or m_t vertices clockwise from the last one marked and put a counter there. The player who is unable to find an empty vertex for his counter to be placed loses. In Berlekamp *et al.* (1982) this is called Kotzig's game.

When t equals 1, that is when a player counts m places from the previously placed counter and puts his mark there, if possible, it has also been called 'she loves me, she loves me not', for an obvious reason.

We look first at this simplest case. If the largest common divisor of m and n equals d, the progress will be blocked after n/d moves. Hence, if n/d is even, the second player wins, if n/d is odd, the first player wins.

This can hardly be called a game. The players have no choice, and the question of skill does not arise. We look therefore at a slightly more interesting case, that of $m_1 = 1$, $m_2 = 2$, and various values of n.

$n = 1$ is not a game. neither is $n = 2$ of any interest.
$n = 3$. The first player wins, whichever m_i the second uses. (Fig. II. 12(a))
$n = 7$. The first player wins in three possible ways, after 3 or after 4 moves. (Fig. II.12(b))

The three cases $n = 1$, 3, or 7 are the only ones in which the first player can win. In all other cases the second player can win. This is shown in a table in Berlekamp *et al.* (1982, p. 482). For instance for $n = 4$, see Fig. II.12(c), for $n = 6$, see Fig. II.12(d).

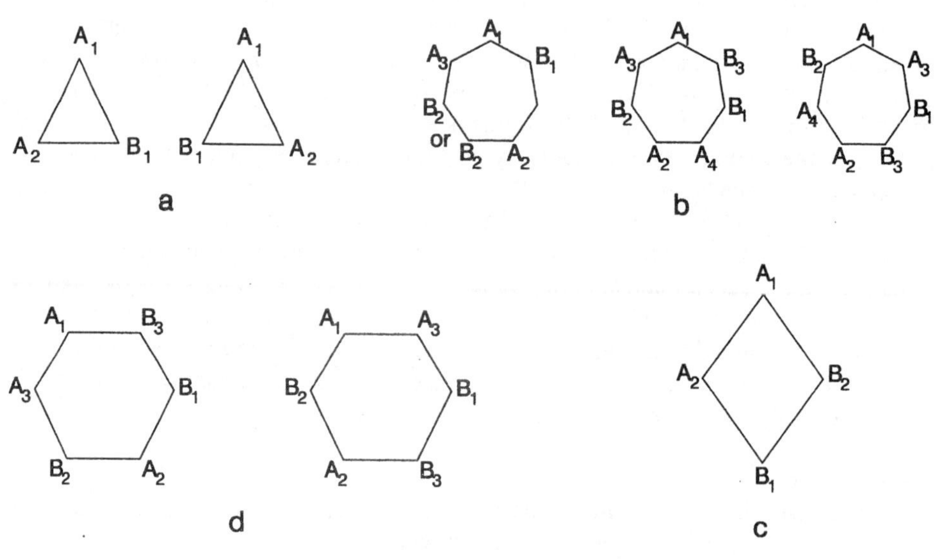

Fig. II.12.

3.

We shall now mention games where a move means joining spots on a plane, subject to various conditions which define the particular game. In all of them two spots are connected by a curve, or a loop is drawn through one single spot. No curve may cross another or itself.

The permitted number of curve terminals at a point will be called the number of lives of that point; when the maximum possible number of curves have touched the point, we declare it 'dead'. A loop takes up two lives of the spot which it touches.

The player who has the last move wins.

Jocasta

There are n spots, each with two lives. Since each curve uses up two lives, the game lasts precisely n moves. If n is odd, the first player wins; if n is even, the first player loses. No player needs to have a strategy, neither can make a mistake. This is hardly a game at all.

More interesting is the game of Sprouts.

Sprouts

Each of n spots has three lives. When a curve has been drawn, a new spot with three lives is placed on it. Two of its lives are already taken up by the curve on which it is drawn, so only one new life is being added.

Each move uses up two lives, and adds one new one. Therefore after m moves $3n - m$ lives will be left. The game will end at the latest when only one life is left, so that

$$3n - m \geq 1 \text{ or } m \leq 3n - 1.$$

No play can last longer than through $3n - 1$ moves. On the other hand, the number of moves must be at least $2n$. We can see this as follows:

After the last move a dead spot cannot be the neighbour of two different live spots, because if it were, these two could be joined without crossing a curve, and the play could continue. Also, each of the live spots left has two dead spots as neighbours. Hence, if the play has lasted for m moves, we have $n + m$ spots, $3n - m$ of them live, $2(3n - m)$ dead which are neighbours of the latter, and other dead spots, say k in number. So

$$n + m = 3(3n - m) + k \text{ or } m = 2n + \tfrac{1}{4}k.$$

It follows that $m \geq 2n$. Incidentally, it also follows that k is a multiple of 4.

Examples

$$n = 1, \quad 3n - 1 = 2, \quad 2n = 2, \quad m = 2 \quad \text{(Fig. II.13(a))}$$

$$n = 2, \quad 3n - 1 = 5, \quad 2n = 4, \quad m = 4 \text{ or } 5 \quad \text{(Fig. II.13(b))}$$

A new spot is black when it appears first.

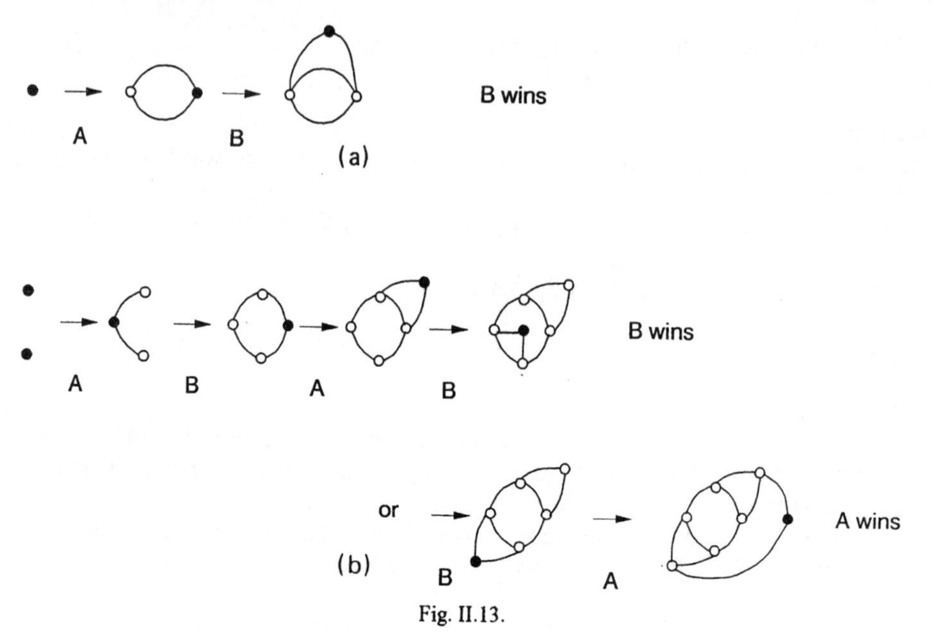

Fig. II.13.

Brussels Sprouts

The rules are the same as for Sprouts, but instead of dots we have crosses, and a curve can start or end at any of the four bars of a cross. Each cross has four lives. After each move we draw a new cross, by adding a crossbar to the new curve.

Because each move deletes as well as adds two lives, the number $4n$ of lives of the originally n crosses remains the same throughout a play. It can be shown that every play lasts precisely for $5n - 2$ moves, so that, again, no mistake can be made and the first (or second) player must win, if n is odd (or even). The proof depends on topological concepts, such as 'inside' or 'outside' of a loop.

We give an example for originally two crosses. The second player wins. (Fig. II.14.)

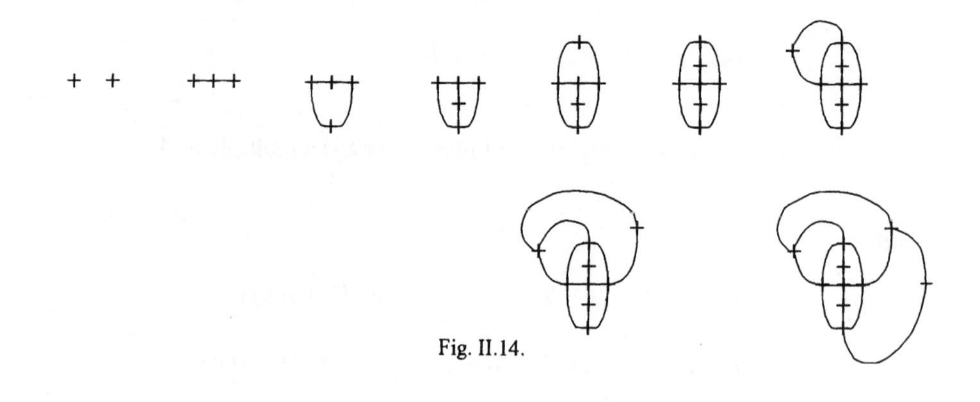

Fig. II.14.

The original two crosses had altogether 8 lives, and 8 lives are left. They appear in 8 different sections of the plane: the play is over.

These, and similar games, some with (as yet) incomplete theories, are mentioned in Berlekamp *et al.* (1982).

Sim

This game (Simmons, 1969) is played on the 15 edges (6 sides, 9 diagonals) of a complete hexagon. The players take turns in drawing one of these edges, with the aim of forcing the opponent to form a triangle of edges drawn by him, thus to lose.

As an introduction, we consider this game to be played on the six edges of a complete square, or on the 10 edges of a complete pentagon.

First, take the square. The six edges can be divided into two triples, neither a triangle. This can be done in various ways, but all of them are topologically equivalent. In any case, two vertices are met by two sides of one triple and one side of the other.

The first player choses an edge, and so does the second player. These two edges can be considered to belong to two different triples. When during the next moves each player keeps to his own triple, he is safe. The play will end in a draw.

Similarly, the 10 edges of the pentagon can be divided into two quintuples, apart from topological equivalence in one single way. Once more, the first respective moves decide which quintuples the players ought to keep to, to avoid a loss. The play will end in a draw.

Now we come to the hexagon. The number of all edges is 15, an odd number, which cannot be equally divided. Results relevant to this case are mentioned in Appendix IV. It follows from Bostwick (1958) that a draw is impossible, and from Goodman (1959) that when the play has ended, with each edge chosen by one or the other player, then at least two triangles will necessarily be completed, that is all three sides will belong to the same player. From Harary (1972) it follows that when 14 edges have been chosen, 7 by each player, without having yet completed a triangle, then the first player, whose turn it now is, is bound to complete a triangle and to lose.

All sets of such 14 moves are topologically equivalent. Consider the situation in Fig. II.15(a) (Gardner, 1973).

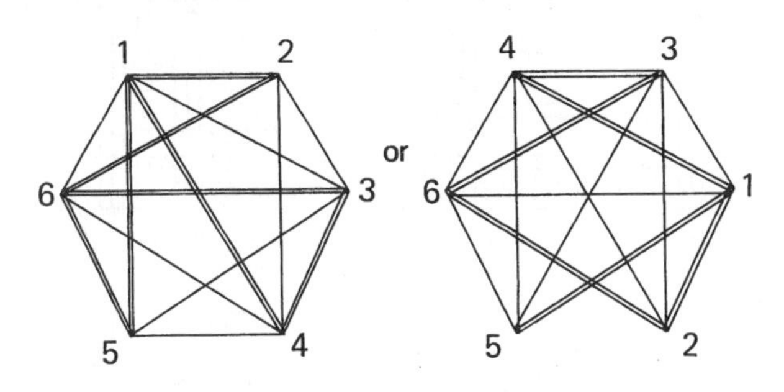

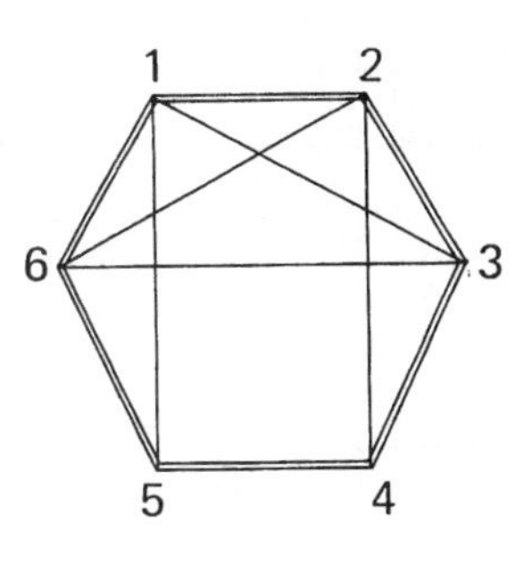

Fig. II.15.

One player has drawn the single lines, the other the double lines. It is the turn of the first player to move. Only the edge 25 is left to be drawn. It will complete two triangles, either 125 and 256, or 235 and 245. In either case the first player loses.

However, this does not mean that the first player must always lose. In Fig. II.15(b) it is the turn of the player who draws single lines, the second player. But he cannot win. To avoid immediate defeat, he draws 14, or 25. The first player then draws the other edge of this pair, and the second must draw either 35, or 46; in either case he loses.

All the first moves of the first player are topologically equivalent. The first move of the second player either connects with the line drawn, or does not connect with it. When, after this, the first player has made his second choice, then the possible further plays lead, in equal numbers, to him winning, or to him losing (Rounds and Yau, 1974). It is impracticable to describe a winning strategy, but the authors give a fairly efficient one, based on the following idea: minimize the number of edges which would have to be removed from the list of still available edges, after one of them has been selected. For example, if in a triangle *abc* the edge *ab* has already been selected, while the list of still available edges contains both *ac* and *bc*, and now *bc* were chosen, then the edge *ac* would have to be removed from that list, because it would complete a triangle.

Cram

The game of Cram (see Gardner, 1974) is played on an assembly of squares. The players cover, alternately, two adjacent squares by a 'domino'. The player who cannot find two adjacent empty squares to cover, loses.

On a *n* by *m* rectangular board, with *m* as well as *n* even, the second player wins, by copying centric symmetrically, the first player's moves. (see Fig. II.16(a)).

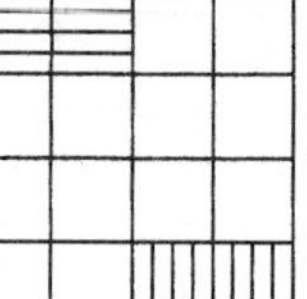

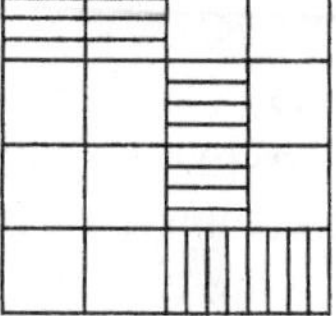

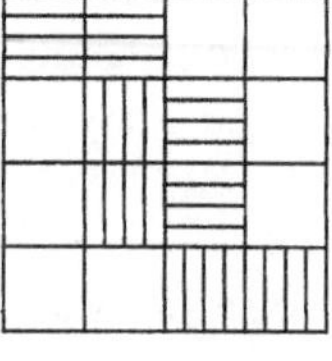

(a)

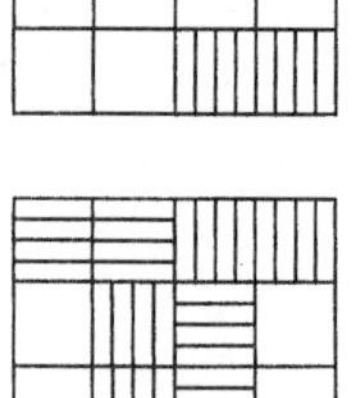

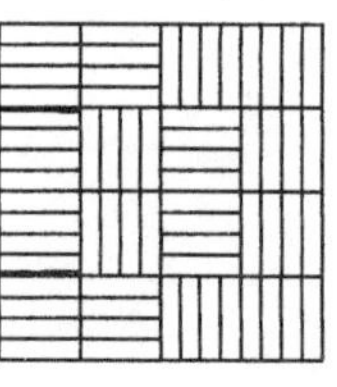

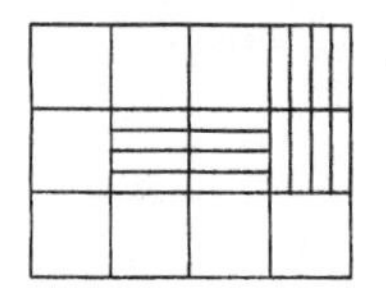

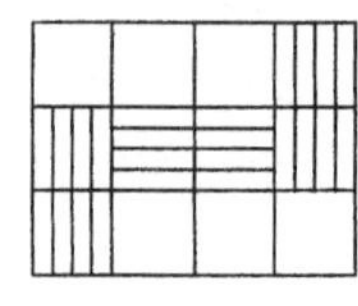

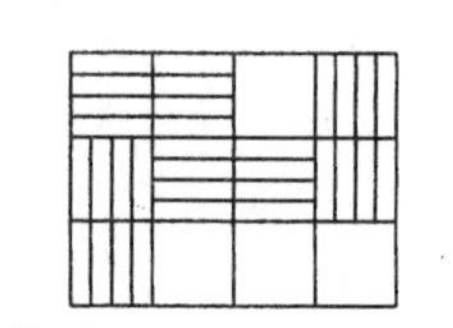

(b)

Fig. II.16.

On a rectangular board with one side even and another side odd, the first player wins by covering the two central squares and afterwards echoing the opponent's moves (Fig. II.16(b)).

4.

The next games we introduce are 'positional'. They are defined by Hales and Jewett (1963) as being played on a finite set X with which is associated a collection $S = (F_1, F_2, ...)$ of subsets of X. Each player claims, in turn, a previously unclaimed element of X.

If one player has claimed every element of some F_i, he has won. If every element has been claimed, but no player has won, the play is a tie.

The most familiar game of this type is Noughts and Crosses. But before we deal with it, we consider more general situations.

A combinatorial game

This is a two-person positional game described by Csirmaz (1980). The players choose alternately m and n elements of X and it is proved that if for $m = n$, $0 < \mu < 1$

$$\frac{m}{n}(\mu^{-m} - 1) \leq 1 \text{ and } \sum_{F_i} \mu^{|F_i|} < \mu^m$$

(where $|F_i|$ is the number of elements in F_i) then the second player can prevent the first from winning: he can at least force a tie. In particular, if $m = n = 1$, $\mu = \frac{1}{2}$, $|F_i| = c$ for all i, then the second player can force a tie if

$$\sum_{F_i} (\tfrac{1}{2})^c < \tfrac{1}{2}.$$

On the other hand, if

$$\sum_{F_i} (\tfrac{1}{2})^c = \tfrac{1}{2} \qquad (*)$$

then the following family allows the first player to win (Erdös and Selfridge, 1973). (This does not mean, though, that whenever (*) holds, the first player can win.)

$X = (a_1, ..., a_n, b_1, ... b_{n-1})$; 2^{n-1} subsets, each with $c = n$, are constructed as follows: each subset contains a_n, and either a_i or b_i $(i = 1, 2, ..., n-1)$, but not both. Then

$$\sum_{F_i} (\tfrac{1}{2})^c = 2^{n-1} \cdot (\tfrac{1}{2})^n = \tfrac{1}{2},$$

For instance, $n = 4$, and the 8 subsets are

$$(a_1,a_2,a_3,a_4),\ (a_1,a_2,b_3,a_4),\ (a_1,b_2,a_3,a_4)\ (a_1,b_2,b_3,a_4),$$

$$(b_1,a_2,a_3,a_4),\ (b_1,a_2,b_3,a_4),\ (b_1,b_2,a_3,a_4)\ (b_1,b_2,b_3,a_4),$$

The first player wins if he proceeds as follows. He starts by choosing a_4. Then, if the second player chooses a_i, he chooses b_i, and vice versa. With each of his moves, the second player can only block half the number of the remaining subsets, and on the fourth move a complete subset has been collected.

It is easy to see that if the game is played with fewer than 2^{n-1} subsets, then the second player can prevent the first from winning.

Three-in-a-row

This familiar game is called Tic-tac-toe in America, Noughts and Crosses in Britain, Eck-meck-steck in German, Boter, Melk, Kaas in Dutch, and no doubt many other names in other languages.

It is played on a 3 by 3 board of squares. The two players put, turn by turn, a O or a X into a yet empty square. The player who fills first a row (vertical, horizontal or diagonal) with his own sign wins. If neither can win, the result is a tie.

Let the first player play Nought. He can put his sign into any of the nine squares, but because of symmetries, there are only three essentially different possibilities, viz.

```
. . .     O . .     . . .
. O .     . . .     O O .
. . .     . . .     . . .
 (A)       (B)       (C)
```

(A) it is easy to see that

```
. . .
. O .
X . .
```

leads to a draw, whichever way Nought continues. While

```
. X .
. O .
. . .
```

allows Nought to win, thus:

```
. X .    . X X    O X X
. O .    . O .    ? O .
O . .    O . .    O . ?
```

and the two squares marked ? cannot be blocked simultaneously.

(B)

```
O . .
. X .
. . .
```

leads to a draw, while in the other cases

```
O X .    O . !    O . .    O . !
. ! .    . . .    . ! X    . . .
. . .    X . .    . . .    . . X
```

Nought can win, by putting his mark into the square with ! in it. Again, we have only mentioned one specimen out of a set of equivalent positions.

```
      . . .     . . .     . . .
(C)   O × .     O . .     O . ×
      . . .     × . .     . . .
```

lead to draws, while in the case

```
. . .
O . .
. . ×
```

Nought can win.

While Cross can prevent Nought from winning he cannot ensure to win himself.

Some games are equivalent to Noughts and Crosses, though it might not appear to be so at a first glance. For instance, cards numbered 1, 2,..., 9 are given. Two players pick up, alternately, one of the cards, and the player who obtains first a total 15 of the numbers on his cards wins. This is precisely Noughts and Crosses, since winning means to have selected the numbers in a horizontal, vertical or diagonal row of a three by three magic square, for instance

```
4 3 8
9 5 1
2 7 6
```

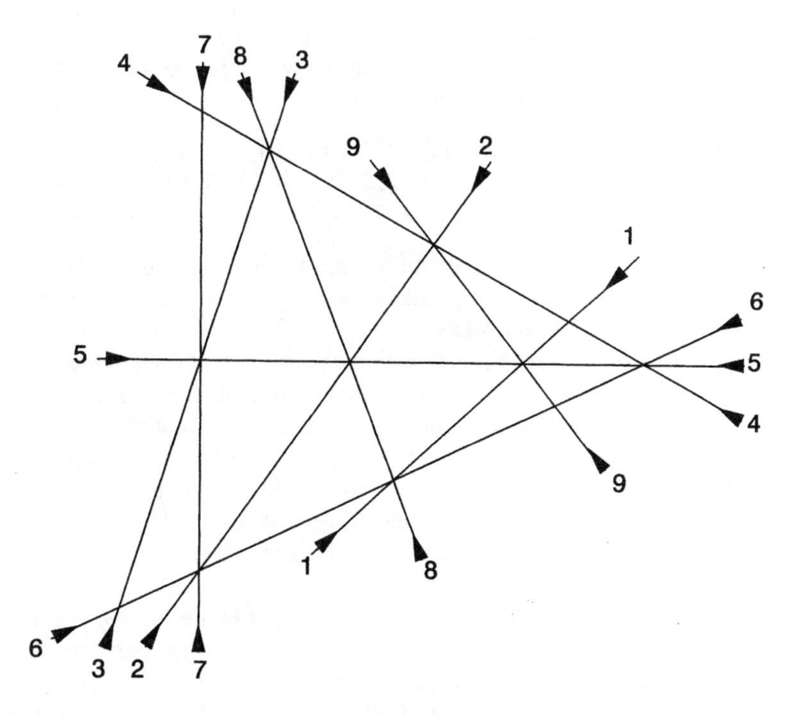

Fig. II.17.

in which the entries in all rows add up to 15.

Another equivalent task is that of finding, out of nine roads, three which pass through the same town on the road map above (Fig. II.17).

It has also been suggested that 'in order to make the game more interesting', restrictions might be imposed such as not allowing either player to occupy the centre at his first move, or at any time except to prevent a row to be completed.

k-in-a-row

It is hardly surprising that after three in a row, we might now be interested in changing 3 to some other number, *k*.

Clearly, when two in a row is played on a two by two board, the player who moves first is bound to win.

We have just seen that when $k=3$, the second player, Cross, can force a draw. This is also true for $k=4$ on a four by four board. This we shall now prove.

Imagine that four in a row proceeds on a 5 by 5 board. For Nought to win, he must start by covering the central 3 by 3 square, or he must put all his four marks outside the centre. The former attempt can be blocked, as we have seen, and the latter is blocked by the opponent entering one of the squares adjacent to the corner squares, horizontally or vertically.

Since Nought can be prevented from winning four in a row on a five by five board, he cannot possibly win on a smaller, four by four board.

We show now that playing *k* in a row on a *k* by *k* board, $k \geq 3$, Cross can always force a draw. We have proved it for $k=3$ or 4, and we proceed to $k=5$.

Let the five squares in the top row of the board be numbered 1, 2, 3, 4, 5 in this order, those in the second row 6 to 10, and so on to 25. To win, a player must cover with his sign one of the following 12 sets of 5 numbers each:

$$1, \underline{2}, 3, \underline{4}, 5;\ 6, \underline{7}, \underline{8}, 9, 10;\ \underline{11}, 12, 13, 14, \underline{15};\ 16, 17, \underline{18}, \underline{19}, 20;$$
$$21, \underline{22}, 23, \underline{24}, 25;\ 1, \underline{6}, 11, \underline{16}, 21;\ 2, 7, \underline{12}, \underline{17}, 22;\ \underline{3}, 8, 13, 18, \underline{23};$$
$$4, \underline{9}, \underline{14}, 19, 24;\ 5, \underline{10}, 15, \underline{20}, 25;\ \underline{1}, 7, 13, 19, \underline{25};\ \underline{5}, 9, 13, 17, \underline{21}.$$

We want to prove that Cross can prevent Nought from achieving this by answering each of Nought's moves by an appropriate blockage. To every square which Nought covers, we must find Cross's immediate answer.

Now if we can find numbers 1,2 ,..., 25 with the property that each of the sets of five above contains a pair, and pairs do not overlap, then if Nought enters a square numbered n_1, Cross ought to enter the square n_2, where (n_1,n_2) is a pair earmarked in the same set. Nought will then never be able to complete a set, that is put five into a row.

The numbers underlined in the list above form such pairs as required.

Scrutiny shows, though, that 13 is not underlined in any of the sets in which it appears. However, if Nought enters the square 13, Cross answers by entering any yet empty square, and later acts as described.

This deals with $k=5$, and further progress in our proof hinges on the possibility of finding suitable pairs in sets of 6, 7,... numbers, when the sets are the numbers of squares in the same (horizontal, vertical or diagonal) row. We find the positive answer in the Theorem of Distinct Represetatives by P. Hall (1935). This theorem states that there exist distinct numbers, one from each of n sets, if and only if the union of any m of the sets ($m \leq n$) contains at least m different numbers. (The necessity is obvious. Hall has proved sufficiency.)

We can use this theorem for our purpose. Take each of the $2k+2$ sets twice. If we find distinct representatives for the resulting $4k+4$ sets, then the representative pairs we are looking for are the two numbers representing identical sets.

Why could we not use this same argument for $k=3$ or 4? Because in these two cases we had 8 and 10 sets respectively, but only 9 and 16 numbers. For $k=5$ we have (with duplicates) $4(k+1)$ sets with k^2 numbers, and k^2-4k-4 is positive when $k \geq 5$.

When we were dealing with $k=4$, we considered a board larger than 4 by 4. This suggests studying generally k in a row on a larger than k by k board. For instance, the Japanese game Go-moku is 5 in a row on a 19 by 19 board. By using a computer, it has been found that k in a row can be won by Nought on a sufficiently large board, for k up to 4. For $k=3$, the board need only be 4 by 4. Nought uses the 2 by 2 centre twice. For $k=4$ we have seen that the board would have to be larger than 5 by 5. On the other hand, for $k \geq 8$, Cross can always force a draw. This is proved in Zetters (1980).

A game called k_n-game has also been investigated (Hales and Jewett, 1963). It is played on a $k \times k \times \ldots \times k$ (n times) hypercube. There are altogether

$$\frac{(k+2)^n - k^n}{2}$$

rows of length k in such a game. The second player can force a draw if $k \geq 3^n - 1$ (for odd k) or $k \geq 2^{n+1} - 2$ (for even k). Also, for each k there exists an n_k such that the first player can win if $n \geq n_k$.

The above inequalities for k were improved in Erdös and Selfridge (1973), but without confirming the conjecture of Hales and Jewett (1963) that when $k \geq 2\,(2^{1/n} - 1)^{-1}$, then the second player can force a draw. This inequality holds when there are more points to be covered than twice the number of rows.

The reader will have noticed that in all these positional games the first player may have a possibility of winning, but never the second player. All the latter can do is force a draw, so in describing the rules it may be said that if he succeeds in this, he has won.

Ovid's Game (Berlekamp *et al.* 1982, p. 672).

As a modification of Noughts and Crosses, we play this game again on a 3 by 3 board, with three counters for either player, but when all counters have been placed, a player slides his own counter into a vertically, horizontally, or diagonally adjacent square. The aim is still to have three in a row.

The first player can win this game if he occupies first the central square. The second player produces

```
× . .       . × .
. O .   or  . O .
. . .       . . .
```

or a position which is equivalent, turned round through 90, 180 or 270°.

The first player moves then

either to the centre of a line not covered by Cross — or to a corner not covered by Cross

```
× . .        . × .
. O .        . O .
. O .        O . .
```

or, again, to an equivalent position.

From now on every move is compulsory (unless you want to lose).

```
×  ×  .        .  ×  ×
.  O  .        .  O  .
.  O  .        O  .  .
   ↓              ↓
×  ×  O        O  ×  ×
.  O  .        .  O  .
.  O  .        O  .  .
   ↓              ↓
×  ×  O        O  ×  ×
.  O  →        ×  O  .
×  O  →.       O  →. →.  .
```

and Nought wins after two more (sliding) moves.

It will be noticed that no diagonal moves were necessary.

Nine Men's Morris

French: Marelle. German: Mühle.

.... the nine men's morris
is fill'd up with mud.

Titania (*Midsummer Night's Dream,* II.ii.)

This well-known game is played with nine counters for each of two players, on a board like that in Fig. II.18. The aim is to place three counters on to the same row (edge). Whenever a player succeeds in this aim, he removes one of his opponent's counters, not in a row of three. When all counters have been placed on the board, a counter is moved along one of the edges to an adjacent point. A player who has only three counters left, may move to any empty point, and the player with only two counters left loses.

If a player can put two of his counters on to two opposite corners of a square which is otherwise empty, then it is too late to prevent him from obtaining three in a row after he has covered one of the other two vertices. At least one of the points between two covered vertices will still be available to him.

Another possibly useful combination is shown in Fig. II.18. It might then be possible to put a counter onto 11 or 20 and then to complete a triplet. Such and similar devices can be frustrated, but they are useful against an inexperienced player.

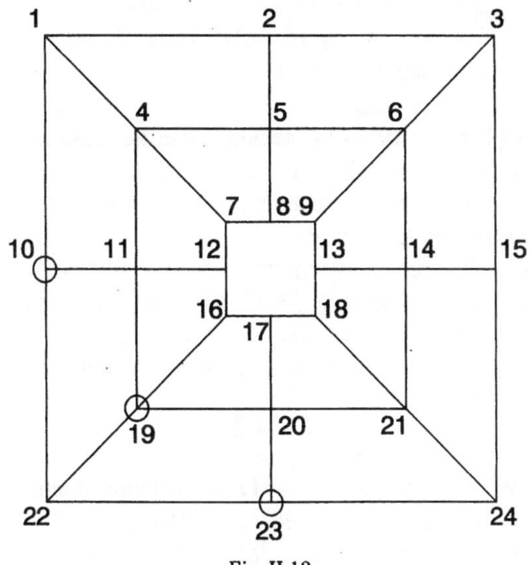

Fig. II.18.

Regretfully, we do not know any winning strategy for this game, but we give an example, with a few typical situations for a player to contemplate.

Assume that after nine moves by both players neither has put three counters into a row, and that the position is as in Fig. II.19(a). W are White's counters, B those of Black, and it is White's turn to move.

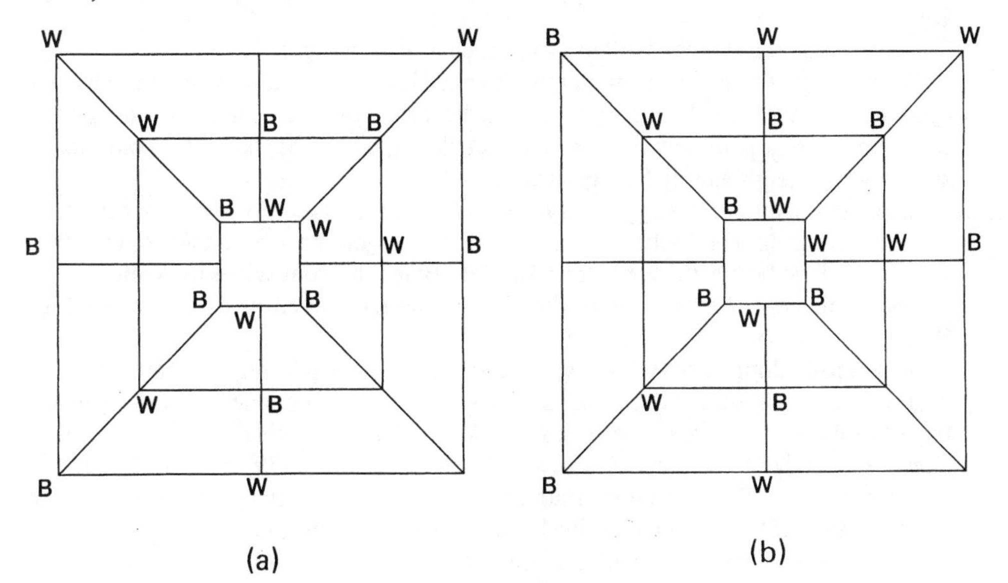

Fig. II.19.

Without any appropriate counteraction, B would win (7, 12, 16) after two moves by 10 → 11, 11 → 12, or (18, 21, 24) by 15 →24, 20 → 21. He cannot move 20 → 21 first, because then W would win (17, 20, 23).

This must be prevented. The first case would be the greater danger. 10 –11 cannot be prevented by 4 → 11, or by 19 → 11, because in either case B would win a counter at once, either (1, 4, 7) by 5 → 4, or (16, 19, 22) by 20 → 19. However, W can prevent (7, 12, 16) by moving 1 → 2. This threatens (1, 2, 3) by 4 → 1, so that B must move 10 → 1, and thereby abandon his aim at (7, 12, 16).

It is now W's move. To prevent B winning a counter by the second possibility, he can move 14 → 21, or 23 → 24, or 9 → 13 which spoils it even more efficiently. Now B cannot make any move towards (18, 21, 24) (see Fig. II. 19(b).

In fact, B can only move 6 → 9, which is pointless, or make one of these moves:

7 → 12, 16 → 12, 22 → 10, 18 → 21.

If B moves 7 → 12, W is free to move 4 → 11 wich would make B reverse to 12 → 7 and W reverse to 11 → 4; nothing would have changed. W could move 8 → 9 or 14 → 21, again without any profit (provided he does not answer B's 15 → 14 by 21 → 24, a foolish mistake). B's moves 16 → 12 or 22 → 10 appear to be equally futile. If B moves 18 → 21, W would move 17 → 18, provoking B's 6 → 9, another impasse, or would move 8 → 9 or 13 → 9, in either case forcing B to answer with 21 → 18.

Here ends our analysis. Must the play end in a draw?

The game with six counters for each player, on the board with only two squares, and no diagonal edges, has been called six men's Morris (Berlekamp *et al.* 1982, p. 672).

Hex

This is played on a rhombus of hexagons, such as that in Fig. II.20.

The four regions outside the rhombus are called Black and White, as shown. The two eponymous players mark, turn by turn, some hexagon with their colour, and the player who constructs a path between the two outside regions of his name by colouring a succession of neighbouring hexagons wins.

An explicit winning strategy is only known for simple cases. However, we can show that the game cannot possibly end in a draw. If all hexagons have been coloured in some way, there must be a path, either from Black to Black, or from White to White.

Assume that each hexagon of the rhombus has been coloured, as for instance in Fig. II.20.

Draw a path along edges which separate a B region from a W region. Start with *a, b, c,* or *d*. We can be sure that arriving at any vertex we can continue, because there are three hexagons at a vertex, one B and one W, and the third either B or W, indicating along which edge to continue. It can also be shown (Gale, 1979) that we shall never return to a vertex. Therefore if we start, say, with *a*, we shall have to finish with *b, c* or *d*. As a matter of fact, we cannot finish with *b,* but this is irrelevant to our proof.

Whenever the path terminates, it indicates a succession of B-hexagons, or of W-hexagons. No tie is possible. In the example B wins; it is even unnecessary to continue until *c*, to see this.

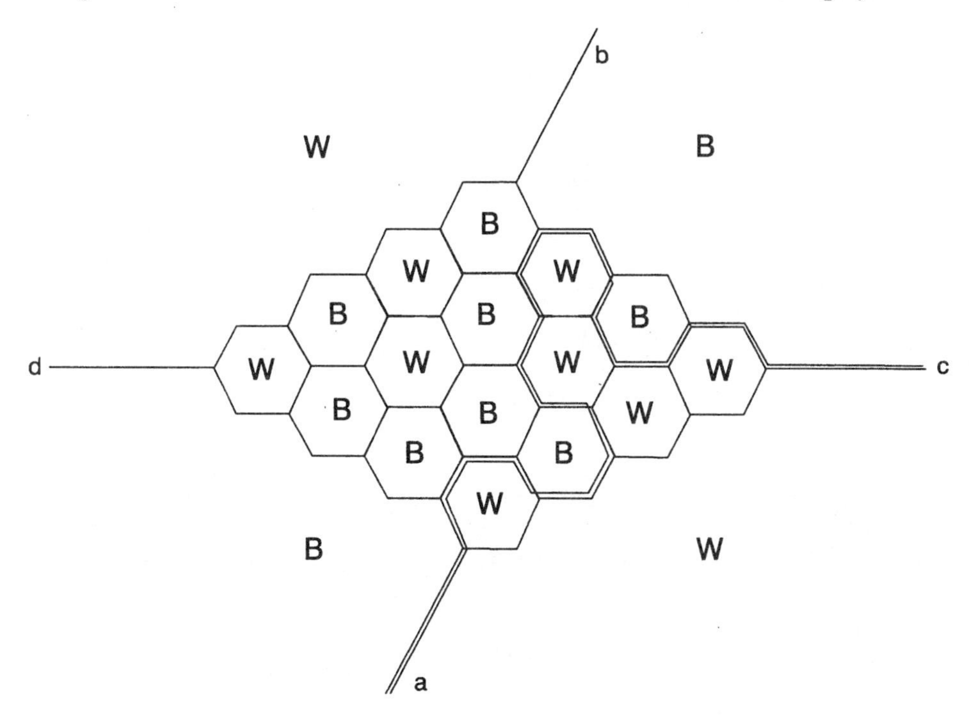

Fig. II.20.

Since no tie is possible, we might ask, who has a winning strategy? It has been proved that it is White, though the identity of this strategy is unknown. What is known is that White must open in an obtuse corner of the rhombus board, otherwise Black has a winning strategy. (Beck *et al.,* 1969).

Reverse Hex is the game on the same Hex-board, but the players try to force the opponent to complete a path between his outside regions. In this version, on an n by n board, the first player can win when n is even, and the second player wins when n is odd (Gardner, 1959, p. 78). According to Evans (1974), in the former case the first player opens in an acute corner.

Some variants of Hex are presented in Evans (1975/6).

Shannon's Game

Two players play on a non-directed graph, on which two vertices are marked. One player, Short, tries to find a path between the marked vertices by enforcing at his moves one of the edges, while the other player, Cut, cuts an edge, to prevent a path of enforced edges being formed.

The analysis of this game requires that we distinguish between the two players, but also between the player who moves first, and the player who moves second. This is readily seen from the following examples (Fig. II.21):

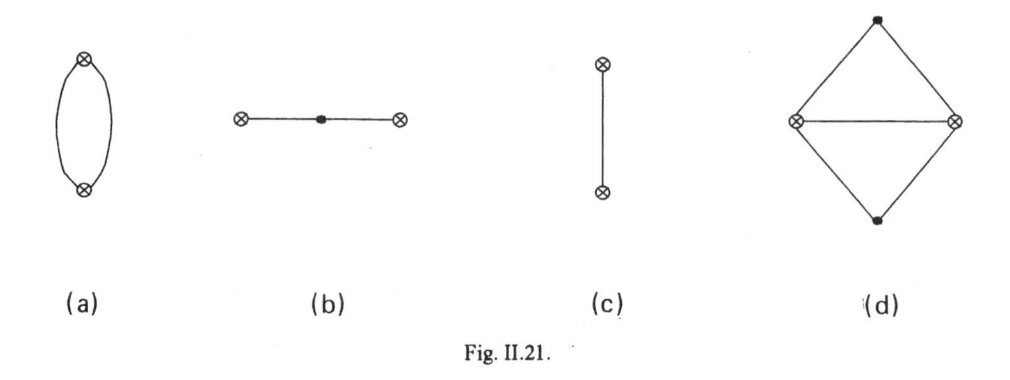

Fig. II.21.

(a) If Short starts, he can ensure that a path exists. Cut cannot do this. Short wins, whether he is the first, or the second player.
(b) If Cut starts, he can cut one edge, and thereby cut the path. If Short starts, he cannot prevent Cut from cutting the not enforced edge. Cut wins, whether he is the first, or the second player.
(c) The first player establishes the path if he is Short, or cuts it if he is Cut. The first player wins, whether he is Short, or Cut.
(d) Once more, the first player wins, whether he is Short, or Cut.

The outcome depends on the graph as well, and on the positions of the marked vertices. More detail will be found in Lehman (1964), who uses some fairly advanced graph theory.

Bridgit

This is the name under which a game invented by D. Gale has been marketed.

Imagine that we superimpose on the graph in Fig. II.21(d) a second graph, whose edges cut those of the first graph. We can then describe a game by saying that each player has his own graph, on which he draws or reinforces edges, and that no two edges may cross. We obtain the game of Bridgit when both graphs are n by $(n + 1)$ lattices, interlaced as in Fig. II.22(a), where n equals 4.

We have only drawn the vertices, and the edges are imagined to connect two vertically or horizontally adjacent vertices. To make this strictly a special case of Shannon's game, we imagine that the bottom four nodes are identical, and so are the top four.

Now Short wishes to connect one of the four black dots at the top and one of those at the bottom by a path of his own edges, while Cut wishes to prevent this by drawing his own edges in a way which connects one of the white dots on the left and one of those on the right, by a path.

The game was invented by D. Gale, and the winning strategy which we describe is due to O. Gross (see Gardner, 1966).

Let Short start. With his winning strategy, he draws first the edge between A and B shown in Fig. II.22(a). Later, whenever Cut crosses the end of the slanting or semi-circular lines also shown, he draws the edge which passes through the other end of that line. (The reader will have no difficulty with smaller or larger boards.)

This strategy does not provide for drawing edges on the margins, but such edges would not contribute anything to the advantage of either player. If a player does move on one of these marginal edges, then the opponent makes any move on his own lattice, and if this move were required later as a part of the strategy described, he makes again an arbitrary move.

It ought to be mentioned, though, that there exist other strategies, which would do just as well.

The final result could be, for instance, that shown in Fig. II.22(b). Short's moves are marked 0, 1a, 2a,..., 11a, and Cut's moves are 1, 2,..., 11.

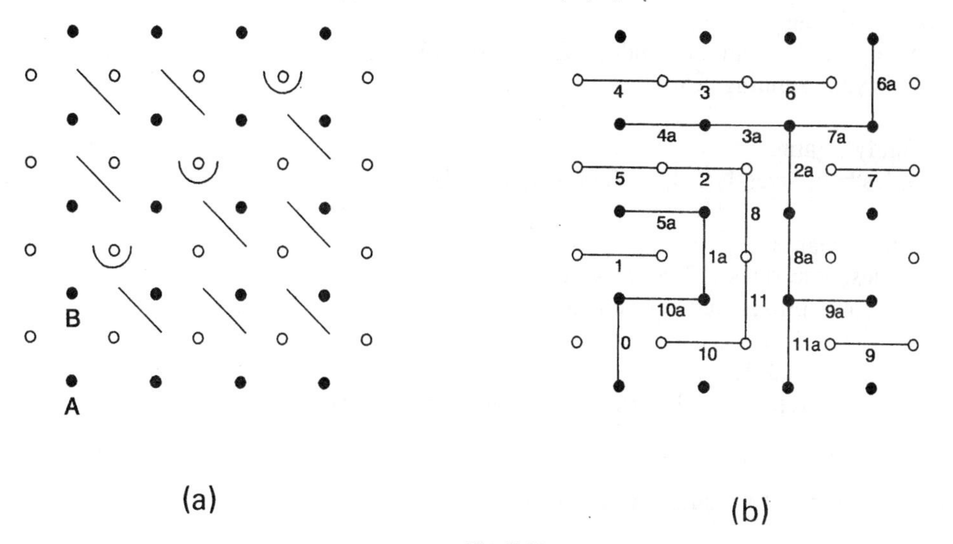

Fig. II.22.

EXERCISES TO CHAPTER II

1. A pile of 14 counters is given. Indicate your winning move, if you are playing
(a) $S(1, 4, 7)$
(b) $S(1, 2, 3, 4, 7)$

2. Fibonacci Nim.
Start with 60 counters. I plays optimally, II removes as many counters as he is allowed to. How does the play proceed?

3. Schwenk's Nim.
$f(n) = 3n$. Start with 30 counters. I plays optimally, II removes as many counters as he is allowed to. How does the play proceed?

4. Nim.
The sizes of three piles are 28, 29, 30. I plays optimally, II takes, if the size of the largest pile is n, the largest integer not exceeding $\frac{1}{2}n$.
How does the play proceed?

5. Welter's Game.
Five piles are given. Their sizes are 2, 3, 6, 7, 9.
Compute your next winning move:
(a) by Sprague's method.
(b) by Welter's 1954 method.

6. Dawson's Chess.
During a play, there are three piles left, with sizes 1, 2, 3.
Which is your winning move?

7. Grundy's game.
Four piles have sizes 1, 2, 4, 6. Your next move is?

8. Lasker's Game.
Three piles, with sizes 1, 3, 5, are given.
State all your options, and your opponent's best answers.

9. Whythoff's Game.
Two piles, with sizes 4 and 5, are given. Your winning move?

10. Knights.
Two piles, with sizes 4 and 5 are given. Your winning move?

11. Moore's game. $k=2$
Start with pile sizes

$$\begin{array}{rr} 1 & 1 \\ 3 & 11 \\ 5 & 101 \\ 7 & 111 \\ 9 & 1001 \\ 11 & \underline{1011} \\ & 2200 \end{array}$$

Describe a play, when A makes always a winning move, and B removes always a whole pile.

12. Divisions.
$N=135$.
Which is your best move?
Indicate all options of your opponent, and your winning replies.

13. Three piles, with sizes 3, 5, 7 are given.
Indicate your winning move, if you are playing
(a) $S(2, 3, 4)$ on each component,
(b) Nim,
(c) Welter's Game,
(d) Kayles,
(e) Grundy's game,
(f) Lasker's Game.

14. Sylver Coinage.
(a) A has called 3, B has called a number of the form $m = 3k + 1$.
List the numbers expressible as weighted sums of 3 and m.
(b) As (a), but m is of the form $3k + 2$.

15. Sylver Coinage (again).
A has called 4, B has called 5. How can A win?

16. Marguerite.
$m_1 = 1, m_2 = 2, n = 5$.
Prove that the second player can win.

17. Ordinary *vs.* Misère-Nim.
Four piles are given, their sizes are 3, 2, 1, 1.
How can A win, playing
(a) Ordinary Nim.
(b) Misère-Nim.

18. Combinatorial game.
(a) Let S be $\{1, 2, 3, 4, 5, 6\}$ and the family of its subsets
$(1, 2) \quad (1, 3, 4) \quad (1, 5, 6)$.
We have
$$(\tfrac{1}{2})^2 + (\tfrac{1}{2})^3 + (\tfrac{1}{2})^3 = \tfrac{1}{2}.$$
Must A win?
(b) We have S as in (a), but the subsets are now
$(1, 2) \quad (1, 3, 4) \quad (1, 2, 5, 6)$.
Must A win?

19. Three-in-a-row (Noughts and Crosses).
Prove that

```
O . .
. O .
× . .
```

leads to a draw.

III

A theory of games

The games which we considered in Chapter II can be described in terms of successive moves, such that at each turn a player knows the choices and outcomes of all preceding moves of both players. They could be described by directed graphs, where a vertex corresponded to a position in the play, and an edge to a legal move from one position to a follower one. Such games are called of 'perfect information'. It can be shown, by induction, that such games always terminate with one player winning, or with a draw. (See, for instance, Blackwell and Girshick, 1954, pp. 17ff.)

In the present chapter we assume that each player decides in advance which move he is going to make at his turn, in any possibly emerging situation. Thus we might say that each player makes just one move, by deciding his strategy. This move is, of course, made without his opponent knowing it; all the two players know is the set of strategies available to either of them. These are, therefore, not perfect-information games, though some might be reducible to such games, as we shall see.

We shall now also allow a player to win an amount dependent on the course of the play, not just to win, to lose or to draw. In the last case we would call the amount won by either player zero.

The values of possible final outcomes of a play are called pay-offs, given in a 'pay-off table'. It exhibits the values a_{ij} for a player R who choses his ith strategy, when the opposing player C has chosen his jth strategy. One player loses as much as the other wins. Such a game is a 'zero-sum' game, or a 'matrix game', because of the way in which the values are displayed.

C chooses column

		1	2	. . .	m
R	1	a_{11}	a_{12}	. . .	a_{1m}
chooses	2	a_{21}	a_{22}	. . .	a_{2m}
row	.	.	.	. . .	.
	.	.	.	. . .	.
	.	.	.	. . .	.
	n	a_{n1}	a_{n2}	. . .	a_{nm}

Either player wishes, naturally, to make that choice which maximizes his gain, or minimizes his loss. But since neither player knows the choice of strategy on which his opponent has decided, such a straightforward choice is impossible, so that some other criterion must be applied. We assume, then, that each player makes that choice of strategy which makes the smallest possible gain, resulting from the opponent's strategy, as large as possible. We may call this a maximin strategy.

As a typical, though somewhat contrived example, consider the following game.

One or Nought Game

R and C choose, independently, one of the numbers 1, 0, or –1. Call R's choice r, and C's choice c. The payment from C to R is then

$$r\,|\,r-c\,|+c(r+c).$$

This rule produces the pay-off table

	–1	0	1	row minima
–1	2	–1	–2	–2
0	1	0	1	0
1	2	1	2	1
column maxima	2	1	2	

Using maximin strategies, the players argue as follows:

If R chooses

$r=-1$	he could lose as much as	2
$=\ 0$	the worst result would be	0
$=\ 1$	he wins at least	1

therefore he chooses $r=1$.

If C chooses

$c=-1$	he might lose as much as	2
$=\ 0$	he might lose	1
$=\ 1$	he might lose	2

therefore he chooses $c=0$.

The result is a payment of 1 from C to R.

If a player now looks at whether he would have made another choice if he had known the strategy of the opponent, he finds that he would have made precisely that choice which he in fact had made. This is so, because the result 1 happens to be the smallest value in its row and also the largest value in its column. We call such a point a 'saddle point'.

Not every pay-off table has a saddle point. Consider, for instance the Number Game.

Number Game

R chooses a pair of numbers from the set (1, 2, 4), and C chooses a single number from the same set. If C has matched one of R's numbers, he receives that amount from R, but loses the amount of the other number. But if C has not chosen either number of R's choice, he loses to R the sum of R's numbers. This is reflected in the pay-off table

	1	2	4	row minima
(1,2)	1	−1	3	−1
(1,4)	3	5	−3	−3
(2,4)	6	2	−2	−2
column maxima	6	5	3	

According to their respective maximin strategies, R would choose (1,2) and C would choose 4. R will be quite happy with this, since he gets 3, more than he could hope for, but C will regret not having selected 2.

Such instability in the effect of the pair of strategies is unavoidable, if the game is played just once. But if the game is played repeatedly, (when a player would be well advised not to repeat his choice every time) then the players might consider choosing their strategies, that is their row or column, with such relative frequencies that in the long run the worst possible average outcome (which depends also on the opponent's choices) is as advantageous as possible. We call this a 'mixed' optimal strategy, while in cases of matrix games with a saddle point, both players choose 'pure' strategies. The result, that is the payment from C to R is called the 'value' of the game.

While the saddlepoint of a matrix is easily identified, if it exists, we need a more sophisticated approach to find mixed optimal strategies. There exist various methods for doing this. We describe here one based on linear programming (see Appendix I).

If R decides to choose row i with frequency s_i, then he obtains

$$\sum_{i=1}^{n} a_{ij} s_i.$$

when C chooses column j. If we write v for

$$\min_j \sum_i a_{ij} s_i$$

then

$$a_{ij}s_1 + \ldots + a_{nj}s_n \geq v \ (j = 1,2,\ldots,m).$$

Also

$$s_i + s_2 + \ldots + s_n = 1, \text{ and } s_i \geq 0 \ (i = 1,2,\ldots,n).$$

The unknown v is to be made as large as possible by the choice of the s_i.

Now let us assume that the maximum of v is positive. Introduce new variables $x_i = s_i/v$. Then $\sum_i x_i$ equals $1/v$, and we have the problem

minimize $x_1 + x_2 + \ldots + x_n$

subject to

$$\sum_{i=1}^{n} a_{ij}s_i \geq 1 \ (j = 1,2,\ldots,m)$$

$$x_i \geq 0(\ i = 1,2,\ldots,n).$$

This is a linear programming problem. When it is solved, then the value of the game is the reciprocal of $\sum_i x_i$, and $s_i = vx_i$.

We have assumed that the maximum of v is positive. This is certainly the case if all a_{ij} are positive. If this condition is not satisfied, then we add a sufficiently large constant c to all a_{ij}, to make $a_{ij} + c$ positive. The optimal strategy remains unaffected, and when we have the final v of the modified problem, then we reduce v by c to obtain the value of the original game.

This was R's solution. Now let us look at that of player C. He chooses $t_1, \ldots, t_m$ $(t_1 + \ldots + t_m = 1)$ so that

$$\max_i \sum_{j=1}^{m} a_{ij}t_j = w, \quad \text{say} \ (i = 1,\ldots,n).$$

Introduce $y_j = t_j/w$, then we must

maximize $y_1 + \ldots + y_m$ (which equals $1/w$)

subject to $\sum_{j=1}^{m} a_{ij}y_j \leq 1 \ (i = ,\ldots, n)$

$$y_j \geq 0 \ (j = 1,\ldots, m).$$

The two problems, that of R and that of C, are 'dual' problems. Linear programming theory tells us that the two optimal values are equal. We also know from this theory that if we solve one of the two problems by the simplex method (explained in Appendix

I), then the final tableau exhibits also the solution of its dual; we need therefore only solve one of the two problems.

The equality of the two optimal values implies that if one of them has obtained his value, the opponent cannot force him to be satisfied with less. As long as one player holds on to his solution, the other has no reason to change his own.

Of course, a matrix game with saddle point can be solved by this method as well. The answer will show that one of the s_is and one of the t_js equals 1, and all other s_is and t_js are zero.

Then either player has a pure strategy. This happens when the game is of prefect information.

After this interpolated relevant theory, we return to the Number Game.

We solve C's problem

$$\text{maximize } x_0 = x_1 + x_2 + x_3$$

$$\text{subject to } \begin{array}{l} x_1 - x_2 + 3x_3 \leq 1 \\ 3x_1 + 5x_2 - 3x_3 \leq 1 \\ 6x_1 + 2x_2 - 2x_3 \leq 1 \\ x_1, x_2, x_3 \geq 0. \end{array}$$

This problem was used as examples 1 and 3 in Appendix I and the final tableau shows that the optimal $1/v$ and $1/w$ are 1, and these are therefore also the values of v and of w.

C should use his second as well as his third column with frequency ½, and never his first column. R should use his first row with frequency 2/3 and his second row with frequency 1/3, but never his third.

The pay-off table contains actually negative values, but it was unnecessary to take any precaution against the value becoming zero or negative, because it is apparent that R can obtain more than 0, for instance by choosing $x_1 = 5/9, x_2 = 3/9, x_3 = 1/9$, (This would give him at least 1/9, less than the optimum 1.)

Matching Game

R and C choose simultaneously one of the numbers 1, 2, 3. If both choose the same number, R pays C that amount. Otherwise C pays R what R had chosen. (This is a simple generalization of the childish game Matching Pennies, where it is obvious that mixed strategies (½,½) should be used.)

The pay-off table is

	1	2	3
1	–1	1	1
2	2	–2	2
3	3	3	–3

There is no saddle point here.

R can achieve a positive result by using (½,¼,¼), so we do not have to make adjustments, in spite of the negative values in the pay-off table.

The simplex method produces the final tableau

	x_6	x_4	x_5	
x_2	1/6	1/2	0	2/3
x_3	0	1/2	1/4	3/4
x_1	1/6	0	1/4	5/12
	1/3	1	1/2	11/6

To obtain the frequencies s_i and t_j, we multiply the x_i and the y_j by 6/11 and obtain, together with $v = 6/11$,

$s_1 = 6/11, s_2 = 3/11, s_3 = 2/11$ for R's frequencies, and
$t_1 = 5/22, t_2 = 4/11, t_3 = 9/22$ for C's frequencies.

Morra

This game, pictured in the Frontispiece of the book, existed in Egypt about four thousand years ago. R and C show simultaneously a fist (i.e. no fingers), or any number of fingers of one hand. At the same time they call a number, which is a guess of the number of fingers shown by the other player. The scores are computed as follows:

—If no player guesses correctly, or if both guess correctly, the score is 0.
—If only one player guesses correctly, then he wins the number he guessed, plus the number shown by himself.

We present a simplified version, where either one or two fingers are shown. Let (a, b) mean showing a fingers, and shouting b. This produces the pay-off table

	(1,1)	(1,2)	(2,1)	(2,2)
(1,1)	0	2	–3	0
(1,2)	–2	0	0	3
(2,1)	3	0	0	–4
(2,2)	0	–3	4	0

Both players have the same opportunities — the pay-off table is skew-symmetric — and it is obvious that the value of the game must be zero. To apply our method of solution, we must add some constant to all entries. Any constant would do, but it is most convenient to make it 1, so that the value of the adjusted game will be 1, and the s_i will equal the x_i, and the t_j will equal the y_j respectively. The problem with 1 added to the entries of the pay-off table is solved in Appendix I (Examples 2 and 4). Either player has optimal strategies

(0, 4/7, 3/7, 0) or (0, 3/5, 2/5, 0)

and their convex combinations $(0, x, 1 - x, 0)$, where $4/7 \leq x \leq 3/5$.

In this chapter of our book we have so far assumed that both players decide on a maximin strategy, and we have shown how such a strategy can be computed. But a maximin strategy is also imaginable if there is no adversary to beat. (If the possible occurrences are, for instance, due to chance, then we might call such a scenario a game against Nature.)

An example of such a situation is betting on a racecourse.

Betting on a racecourse

The bookmaker B offers to the punter P odds $o_i : 1$ on the horse H_i winning. In other words, if P stakes x_i on H_i, and this horse wins, then P receives $a_i x_i = (o_i + 1)\, x_i$ (because the punter also gets his stake back). Of course, B does not know which horse will win (we hope), so he is not a player, only P is.

P has a total amount 1, say, to stake, and he has to decide how much to bet on one, or on more, perhaps on all the horses. He wishes to use a maximin strategy.

Algebraically speaking, he wishes to maximize his receipt v, where v is the minimum receipt, which means subject to

$$a_i x_i \geq v\ (i = 1, 2, \ldots n) \text{ and } x_i + \ldots + x_n = 1,\ all\ x_i \geq 0. \qquad (*)$$

We write the inequalities $a_i x_i - b_i = v$ $(i = 1, \ldots, n)$, $b_i \geq 0$, that is

$$x_i = (v + b_i)/a_i \quad (i = 1, 2, \ldots n).$$

The equation $x_1 + \ldots + x_n = 1$ gives

$$v = (1 - \sum_i b_i/a_i)/\sum_i (1/a_i).$$

All b_i are non-negative, so that v is largest if they are all 0. This gives

$$x_i = (1/a_i)/\sum_{i=1}^{n} (1/a_i), \text{ and}$$

$$v = 1/\sum_{i=1}^{m} (1/a_i).$$

If this is the way in which P distributes his stake, he receives the same amount whichever horse wins. With any other distribution he would win less.

P will, of course, only act in the manner suggested if v, the amount he receives, exceeds the total of money staked, that is if

$$1/\sum_i (1/a_i) > 1, \text{ or } \sum_i 1/a_i < 1.$$

This criterion can be derived without computing the required distribution of the x_i.

A special case of the so-called Lemma of Farkas (1901–2) states that if all solutions of the system of homogeneous inequalities

$$a_{11}x_1 + .. + a_{1n}x_n \geq 0$$

$$\vdots$$

$$a_{m1}x_1 + .. + a_{mn}x_n \geq 0$$

$$x_i \geq 0 \ \ (i = ,\ldots,n)$$

satisfy also

$$b_1x_1 + \ldots + b_nx_n \geq 0$$

then there exist non-negative y_j ($j = 1,..., m$) such that

$$a_{11}y_1 + .. + a_{m1}\, y_m \leq b_1$$

$$\vdots$$

$$a_{1n}\, y_1 + \ldots + a_{mn}y_m \leq b_m \, .$$

In order to use this theorem, we observe that the receipt of P will not exceed 1, if all solutions of the conditions (*) above satisfy also $v < 1$, which can be written

$$x_1 + .. + x_n - v \geq 0.$$

According to Farkas's Lemma, we see that:

If all solutions of

$$a_{i}x_i - v \geq 0$$
$$x_i, \ v \geq 0$$

satisfy also

$$x_1 + .. + x_n - v \geq 0.$$

then there exist y_j such that

$$a_j\, y_j \leq 1 \ \ (j = 1,\ldots,m)$$
$$-(y_1 + .. + y_{m)} \leq -1$$
$$y_j \geq 0$$

This is the case when

$$\sum_j 1/a_j \geq \sum y_j \geq 1.$$

If this holds, the maximum of v will not exceed 1, which is the same as saying that this maximum will exceed 1 if $\sum_j 1/a_j < 1$, in conformity with the condition found above.

Examples

a_1	$1/a_i$	x_i	gain	net gain
3:1	1/3	4/9	4/3	1/3
4:1	1/4	3/9	4/3	1/3
6:1	1/6	2/9	4/3	1/3
	3/4			

N.B. Do not expect that these odds will ever be offered.

a_i	$1/a_i$	x_i	gain	net gain
3:2	2/3	8/13	12/13	–1/13
4:1	1/4	3/13	12/13	–1/13
6:1	1/6	2/13	12/13	–1/13
	13/12			

N.B. Do not bet!

To round off this chapter, we add a few remarks about non-zero-sum games.

The Prisoner's Dilemma

Two prisoners were caught with stolen goods in their possession, and held incommunicado. They are approached by an intermediary (ignore the immorality of such goings on) with the following proposition:

—If you plead guilty, but the other fellow does not, you receive £10. The other fellow will be fined £20.
—If you both plead guilty, you will both be fined £10.
—If neither of you pleads guilty, both will go free. However, there will be no gratuity!.

If the two prisoners have heard about the theory of games, they might construct the following pay-off table:

	My pal pleads guilty B_1	My pal does not plead guilty B_2
A_1 I plead guilty	−10	10
A_2 I plead not guilty	−20	0

Each notices a saddle point at (A_1, B_1) and will plead guilty; each will be fined £10.

If we think of an equilibrium pair as a pair of strategies such that neither player has an incentive to change his own strategy as long as the other sticks to his, then (A_1, B_1) is an equilibrium pair as well as a maximin pair.

Now we combine the two pay-off tables, thus

	B_1	B_2
A_1	−10,−10	10,−20
A_2	−20,10	0,0

Concerning the combined pay-off to both, it would have been better if they could have co-operated and both had opted for not pleading guilty.

A detailed study of non-cooperative finite games, of which the non-zero sum game Prisoner's Dilemma is an example, will be found in Alexander (1990), which deals also with *n*-person games, were *n* may exceed 2.

EXERCISES TO CHAPTER III

1. The pay-off table for the well-known game of Paper–Scissors–Stone is

	P	Sc	St
P	0	−1	1
Sc	1	0	−1
St	−1	1	0

Find the optimal strategies of the two players.

2. A has the choice of playing either of the two games against B. The two pay-off tables are

(a)

	B	
	5	−1
A	800	−5

(b)

		B	
	0	2	−5
A	−2	0	3
	5	−3	0

Which game should he choose?

3. Maximize v, subject to

$$x_1 + x_2 \leq 1$$

$$v - a_1 x_1 \leq 0$$

$$v - a_2 x_2 \leq 0$$

$$x_1,\ x_2,\ v \geq 0.$$

4. Suppose bookmakers offered you the following odds on four horses (no other horses run)

H_1	Red Heat	6:1
H_2	Black Jack	7:2
H_3	Yellow Peril	2:1
H_4	White Wash	5:2

You want to bet altogether 1 unit on some or all the horses. Which is your maximin strategy?

IV

Tournaments and rankings

In this chapter we look at round robin tournaments of n players, where each player plays every other.

1.

At the end of the tournament, the players are ranked according to their results; the more games a player has won, the higher ought to be his ranking. However, with this simplistic approach, we run into difficulties.

We exhibit the results in a two-way table, where we enter 1 in row i and column j, if player P_i has beaten player P_j, 0 if P_j has beaten P_i, and ½ if they drew their game. In the diagonal of the table we write 0.

Example

	M				**R = M·E**
$j=$	1	2	3	4	row sums
$i=1$	0	½	½	1	2
2	½	0	0	1	1½
3	½	1	0	0	1½
4	0	0	1	0	1

We shall readily agree that P_1 has won the tournament. But who is second? Have P_2 and P_3 done equally well?

P_2 has only beaten P_4, who was rather weak anyway, while P_3 has won against P_2, his closest rival. We might judge that, on the whole P_3 has outplayed P_2.

Such considerations have led Wei (1952) to suggest that the scores of P_i, n_{ij} say, should each be multiplied by the row totals $\sum_j n_{ij}$, obtained by P_j. Thus the value of a win against a stronger player is more valuable than that against a weaker player, where the strength of a player is assessed from his results in the tournament.

In our example we obtain then

	M·R				**M·R·E**
	1	2	3	4	row sums
1	0	¾	¾	1	2½
2	1	0	0	1	2
3	1	1½	0	0	2½
4	0	0	1½	0	1½

The tie between P_2 and P_3 has been broken, but another has emerged.

Algebraically, we have multiplied a matrix **M** by a vector $\mathbf{E} = (1,1,...,1)^T$ to obtain the vector **R** of the row sums. Then we multiplied **M** by the vector **R** for adjustment. Since $\mathbf{M}\cdot\mathbf{R} = \mathbf{M}^2\cdot\mathbf{E}$, we have, in effect, computed the row sums of $\mathbf{M}^2$.

If we continue this procedure, perhaps to break other, new or still remaining ties, then the row sums will increase, and we must normalize the scores, perhaps by dividing them by the square root of the sum of their squares. The vector of the row sums will converge to an eigenvector corresponding to the largest eigenvalue of **M** (Wilkinson, 1965, pp. 571/2).

Of course, we cannot always break all ties. If, to begin with, all row sums are equal, then this remains so for any power of **M**.

Goddard (1983) suggests the following method for breaking ties. Multiply each n_{ij} by the sum of the ith and the jth row total and compute the row totals of the new matrix. In our example this would lead to

	1	2	3	4	row sums
1	0	1¾	1¾	3	6½
2	1¾	0	0	2½	4¼
3	1¾	3	0	0	4¾
4	0	0	2½	0	2½

with the break of the tie we would expect.

Ali *et al.* (1986) deal with another possible defect in ranking methods. It will be thought to be illogical for a player to rank lower than another player, whom he defeated. Such a 'violation' appeared in our example, concerning P_2 and P_3.

Their method of adjusting scores, which minimizes the number of violations, depends on methods of mathematical programming, which we cannot discuss here. In our example the smallest number of violations, namely 1, occurs when we rank the players in the order 1, 3, 2, 4 or 1, 4, 3, 2, but not every reader would accept that these ratings, in particular the second, have much to recommend them.

Some organizers of tournaments adopt the simplest possible method of breaking ties. If P_i and P_j tie when all results are listed, then they decide that P_i ranks higher, if in the game P_i vs. P_j it was P_i who won. This method would have solved satisfactorily the dilemma in our example, but it does not always produce the same result as Wei's method, even if there are no draws in the tournament (e.g. in Tennis). This is convincingly shown in the example of a tournament of eight players, with the following results, none of which is a draw:

	P_1	P_2	P_3	P_4	P_5	P_6	P_7	P_8	Row totals	Wei scores
P_1	0	1	0	0	1	1	1	1	5	15
P_2	0	0	1	1	1	1	1	0	5	16
P_3	1	0	0	1	1	0	0	1	4	14
P_4	1	0	0	0	0	1	1	1	4	12
P_5	0	0	0	1	0	0	1	1	3	8
P_6	0	0	1	0	1	0	0	1	3	9
P_7	0	0	1	0	0	1	0	0	2	7
P_8	0	1	0	0	0	0	1	0	2	7.

P_1 beat P_2, but Wei's method gives preference to P_2
P_3 beat P_4, and Wei's method gives also preference to P_3
P_6 beat P_5, and Wei's method gives also preference to P_6
P_8 beat P_7, but Wei's method does not break the tie.

The game P_4 vs. P_5 was won by the latter, but P_4's win against the apparently strong player P_1 has the decisive effect on their respective rankings.

2.

In the previous section we considered the row sums of individual results in round robin tournaments, for the purpose of ranking the players. It seems now natural to enquire which sets of row totals are possible in such a tournament.

It is clear that if there are n players, the total of all the row sums must be $\frac{1}{2}n(n-1)$, because this is the number of games played, and each play produces either $1 + 0$, or $\frac{1}{2} + \frac{1}{2}$. It is this total of $\frac{1}{2}n(n-1)$ which is being partitioned into n portions, but not every partition can emerge. For instance, it is not possible for two players to have accumulated $n-1$, because no two players can have defeated all their opponents, since they have also played one another. Equally, it is impossible for two players to have lost all their games, both accumulating 0.

An answer to the question of which individual scores (row sums) are possible is given by Landau (1953) for tournaments where draws are not possible.

A sequence of non-negative integers $s_1 \leq s_2 \leq \ldots \leq s_n$ can be the sequence of scores obtained by two players if and only if

$$\sum_{i=1}^{k} s_i \geq \tfrac{1}{2}\, k\,(k-1) \quad (k = 1, 2, \ldots, n)$$

with equality for $k = n$.

Proof of necessity. Equality for $k = n$ is obvious.

For $k < n$, consider the sub-tournament of any k players. Their total score from the sub-tournament will be $\frac{1}{2}k\,(k-1)$. But their scores from the entire tournament may be larger, since some or all might have won games against opponents not among the k players of the sub-tournament.

Moon (1963) has pointed out that this proof applies unaltered when the points of a game may be split in any way, for instance $\frac{1}{2} + \frac{1}{2}$ for draws.

The proofs of sufficiency were also given by Landau (1953) and Moon (1963) for the respective cases which they considered. These proofs are more involved than those for necessity, and we refer the reader to the original publications.

EXERCISES TO CHAPTER IV

In a round robin tournament, the following scores were obtained by the five players:

$$3, 3, 2, 1, 1.$$

Show how this might have happended, by constructing a win–lose matrix with scores

(a) 1 for a win, 0 for a loss, no draws
(b) 1 for a win, 0 for a loss, $\frac{1}{2}$ for a draw.

Show also Wei adjustments to break ties.

Solutions to exercises

CHAPTER I

1. (a) Exchange 6–16, 11–16, 3–16, 14–16, 2–16, 13–16, 8–16, 6–16.
   (b) Impossible; the two configurations have different parities.
2. Repeat the moves of the example in the text, in reversed order. (cf. Theorem A in the section on Peg Solitaire.)
3. (a) 46–44, 65–46, 44–46, 47–45, 57–55, 45–65, 75–55, 64–44, 52–54, 44–64, 74–54, 73–53, 54–52, 51–53, 42–44, 23–43, 44–42, 41–43, 31–33, 43–23, 13–33, 24–44, 36–34, 44–24, 14–34, 15–35, 34–36, 37–35.
   (b) Use combination A to remove
   52,53,54, 35,45,55, 32,33,34, 23,24,25 36,46,56, 63,64,65, then 42–44.
4. (a) Switch on, in any order 2,4,6,8 (precisely those whose parities you wish to change).
   (b) Switch on, in any order, 1,2,3,4,5,6,7,8,9 (‘symmetric’ case to that in the text).
   (c1) Switch on 5.
   (c2) Switch on 2,4,5,6,8 (‘symmetric’ cases).
   (d1) Switch on 1,9
   (d2) Switch on 1,2,4,6,8,9 (‘symmetric’ cases).

CHAPTER II

1. (a) The sequence of Grundy numbers is periodic, with period 0 1 0 1 2 0 1 2. Take 1.
   (b) The sequence of Grundy numbers is periodic, with period 0 1 2 3 4 . Take 4
2. $$60 \underset{A}{-} 55 \underset{B}{-} 45 \underset{A}{-} 42 \underset{B}{-} 36 \underset{A}{-} 34 \underset{B}{-} 30 \underset{A}{-} 29 \underset{B}{-} 27 \underset{A}{-} 26 \underset{B}{-} 24 \underset{A}{-} 21 \underset{B}{-}$$
   $$15 \underset{A}{-} 13 \underset{B}{-} 9 \underset{A}{-} 8 \underset{B}{-} 6 \underset{A}{-} 5 \underset{B}{-} 3 \underset{A}{-} 0.\ \text{A wins.}$$
3. $$30 \underset{A}{-} 29 \underset{B}{-} 26 \underset{A}{-} 25 \underset{B}{-} 22 \underset{A}{-} 21 \underset{B}{-} 18 \underset{A}{-} 15 \underset{B}{-} 6 \underset{A}{-} 0.\ \text{A wins.}$$

4.

$$\begin{array}{ccccccccccccccccc}
30 & & 1 & & 1 & & 1 & & 1 & & 1 & & 1 & & 1 & & 1 \\
29 & \rightarrow & 29 & \rightarrow & 15 & \rightarrow & 15 & \rightarrow & 8 & \rightarrow & 8 & \rightarrow & 8 & \rightarrow & 4 & \rightarrow & 4 \\
28 & \text{A} & 28 & \text{B} & 28 & \text{A} & 14 & \text{B} & 14 & \text{A} & 9 & \text{B} & 5 & \text{A} & 5 & \text{B} & 3
\end{array}$$

$$\begin{array}{cccccccccccccc}
 & 1 & & 1 & & & & & & & & & & \\
\rightarrow & 2 & \rightarrow & 2 & \rightarrow & 2 & \rightarrow & 1 & \rightarrow & 1 & \rightarrow & & \rightarrow & \\
\text{A} & 3 & \text{B} & 2 & \text{A} & 2 & \text{B} & 2 & \text{A} & 1 & \text{B} & 1 & \text{A} & 0.
\end{array}$$

5. (a) $A{\cdot}B{\cdot}D{\cdot}ABD{\cdot}BCD = BCD$

reduce to 1 by $BCD \rightarrow 1$, that is

$2,3,6,7,9 \rightarrow 1,2,3,6,7.$

(b) $\varphi(a,b,ac,bc,d) = a\varphi(1,ab,c,abc,ad) \rightarrow 1,2,3,6,7$

$\varphi(1,a,b,ac,bc) = \varphi(1,a,c,ac) = \varphi(1,ab,c) = \varphi(b,ab) = \varphi(1,a) = \varphi(1) = 1.$

6. $\begin{array}{c} 1 \\ 1 \\ \underline{11} \\ \overline{11} \end{array}$ $\quad 1,2,3 \rightarrow 1,2.$

7. $\begin{array}{c} 0 \\ 0 \\ 0 \\ \underline{1} \\ 1 \end{array} \rightarrow \begin{array}{c} 0 \\ 0 \\ 0 \hat{+} 1 \\ \underline{1} \\ 0 \end{array}$ i.e. $\begin{array}{c} 1 \\ 2 \\ 1,3 \\ 6 \end{array}$

8. $\begin{array}{c} 1 \\ 100 \\ 101 \\ 000 \end{array}$ The opponent can win.

Options: (3,5) (1,2,5) (1,1,5) (1,5) (1,3,4) (1,3,3) (1,3,2)
Answers: (3,3) (1,2,4) (0,1,1) (1,1) (1,3,1,3) (0,3,3) (1,1,2,2)
Options: (1,3,1) (1,3) (1,1,2,5) (1,3,1,4) (1,3,2,3)
Answers: (0,1,1) (1,1) (1,1,2,2) (1,3,1,3) (1,3,1,3).

9. $\rightarrow 3,5$ or $\rightarrow 1,2$

10 $\rightarrow 2,6.$

11.

$$\underset{\text{A}}{\rightarrow} \begin{array}{cc} 1 & 1 \\ 3 & 11 \\ 5 & 101 \\ 7 & 111 \\ 7 & 111 \\ 1 & \underline{1} \\ & 000 \end{array} \qquad \underset{\text{B}}{\rightarrow} \begin{array}{cc} 3 & 11 \\ 5 & 101 \\ 7 & 111 \\ 7 & 111 \\ 1 & \underline{1} \\ & 002 \end{array}$$

$$\underset{\text{A}}{\rightarrow} \begin{array}{cc} 3 & 11 \\ 5 & 101 \\ 7 & 111 \\ 6 & \underline{110} \\ & 000 \end{array} \qquad \underset{\text{B}}{\rightarrow} \begin{array}{cc} 5 & 101 \\ 7 & 111 \\ 6 & \underline{110} \\ & 022 \end{array}$$

$$
\begin{array}{rll}
 & 5 & 101 \\
\rightarrow & 5 & 101 \\
\text{A} & 5 & 101 \\
 & & 000
\end{array}
\approx
\begin{array}{rll}
\rightarrow & 5 & 101 \\
\text{B} & 5 & 101 \\
 & & 202
\end{array}
\approx
\begin{array}{c}
\rightarrow \\
\text{A} \\
\text{wins}
\end{array}
\approx
\begin{array}{c}
0 \\
0 \\
0
\end{array}
$$

12. Delete 1
    Options: 3 5 15 (with 3,5) 9 (with 3)
    Answer: 5 3 9 15
    Options: 45 (with 5,9,15) 27 (with 3,9)
    Answer: 27 45
13. (a) → 3,3,7
    (b) → 2,5,7 or 3,4,7 or 3,5,6
    (c) as for Nim (case (b))
    (d) → 3,4,7
    (e) → 3,2,3,7
    (f) → 3,5,1.
14. (a) Numbers of the form
    $3t = 3t$
    $3t + 1 = 3(t - k) + m \quad (t \geq k)$
    $3t + 2 = 3(t - 2k) + 2m \quad (t \geq 2k)$
    (b) Numbers of the form
    $3t = 3t$
    $3t + 1 = 3(t - 2k - 1) + 2m \quad (t \geq 2k + 1)$
    $3t + 2 = 3(t - k) + m \quad (t \geq k)$
15. So far covered 4,5,8,9,10,12ff.
    A calls 11.
    If B calls 2, A calls 3 and wins.
    If B calls 3, A calls 2 and wins.
    If B calls 6, A calls 7, B calls 2 or 3 and loses.
    If B calls 7, A calls 6, B calls 2 or 3 and loses.
16. Let the vertices of the pentagon be numbered 1,2,3,4,5, in clockwise order.
    A_1 in 1, then B_1 in 2.
    A_2 in 3, or 4, B_2 in 5. (wins).
17. (a) (3,2,1,1) → (2,2,1,1) A wins.
    (b) (3,2,1,1) → (2,2,1,1)

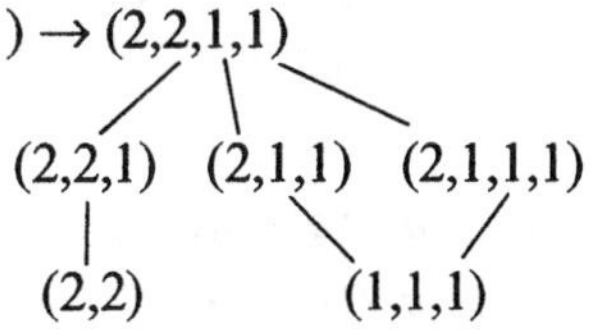

```
              (2,2,1,1)
            /     |     \
    (2,2,1)   (2,1,1)   (2,1,1,1)
       |           \       /
     (2,2)         (1,1,1)
```

A wins.

18. (a) No!
    A must start with 1, B answers with 2.

If now A chooses 3 or 4 or 5 or 6
B answers with 4 3 6 5
and at his next move B prevents A from completing a subset.

(b) $(\frac{1}{2})^2 + (\frac{1}{2})^3 + (\frac{1}{2})^4 < \frac{1}{2}$.
B can prevent A from winning, just as in (a).

19.
```
O . .       O . .       O × .
. O .   →   . O .   →   . O .
× . ×       × O ×       × O ×
```
then either
```
O × .       O × .       O × O
. O O   →   × O O   →   × O O
× O ×       × O ×       × O ×
```
or
```
O × .       O × .       O O ×
O O .   →   O O ×   →   × O O
× O ×       × O ×       × O ×
```

CHAPTER III

1.

	x_1	x_2	x_3	
x_4	1*	0	0	1
x_5	0	1	0	1
x_6	0	0	1	1
	−1	−1	−1	0

Final tableau

	x_4	x_5	x_6	
x_1	1	0	0	1
x_2	0	1	0	1
x_3	0	0	1	1
	1	1	1	3

$x_i = 1/3$ $(i = 1,2,3)$ Obvious!

2. Game (a) has a saddle point with value; (b) has a skew-symmetric pay-off matrix, hence value 0. Therefore A should choose game (b). Strategies for both players: 3/10 : 1/2 : 1/5.

3. Final tableau

	b_1	b_2	b_3	
x_1	$a_2/(a_1+a_2)$	$-1/(a_1+a_2)$	$1/(a_1+a_2)$	$a_2/(a_1+a_2)$
x_2	$a_1/(a_1+a_2)$	$1/(a_1+a_2)$	$-1/(a_1+a_2)$	$a_1/(a_1+a_2)$
v	$a_1a_2/(a_1+a_2)$	$a_2/(a_1+a_2)$	$a_1/(a_1+a_2)$	$a_1a_2/(a_1+a_2)$
	$a_1a_2/(a_1+a_2)$	$a_2/(a_1+a_2)$	$a_1/(a_1+a_2)$	$a_1a_2/(a_1+a_2)$

Observe that $a_1a_2/(a_1+a_2) = 1/(1/a_1 + 1/a_2)$.

This problem is the race-course problem of the text, for two horses.

2.

	$1/a_i$	x_i
H_1	1/7	9/62
H_2	2/9	14/62
H_3	1/3	21/62
H_4	2/7	18/62
	62/63	1

You are being paid (including your stake on the winning horse) 63/62, whichever horse wins: you win 1/62.

CHAPTER IV

(a)

					Wei scores
0	0	1	1	1	4
1	0	1	1	0	6
0	0	0	1	1	2
0	0	0	0	1	1
0	1	0	0	0	3

(b)

					Wei scores
0	1	½	1	½	5½
0	0	1	1	1	4
½	0	0	½	1	3
0	0	½	0	½	1½
½	0	0	½	0	2

Appendix I: Linear programming and the theory of games

Linear programming deals with the problem of maximizing or minimizing a linear function of non-negative variables, subject to linear constraints, which are equations or inequalities.

There is no essential difference between maximizing and minimizing, since max $f(x)$ equals $-\min(-f(x))$. Also, an inequality

$a_1x_1 + \ldots + a_n x_n \le c$ can be written as an equation

$a_1x_1 + \ldots + a_n x_n + x_{n+1} = c$, with $x_{n+1} \ge 0$. We call x_{n+1} a 'slack variable'. Conversely, an equation can be written as a pair of inequalities, in an obvious way.

We shall describe the theory and the practice of linear programming to the extent to which it is relevant to the theory of games (for which see von Neumann (1928) and also von Neumann and Morgenstern (1953)). We shall also assume that the conditions are not contradictory, and that the solution is finite. For more detail see, for instance, Vajda (1981).

Various algorithms have been devised for solving linear programmes. For our present purpose the most convenient of these algorithms is the simplex method of G. B. Dantzig.

We start by changing the m inequalities in n variables to equations in $n + m$ variables by adding a slack variable in each of the inequalities.

The function to be maximized is the 'objective function'. The algorithm is iterative. At each step, we set n variables equal to 0, so that the m equations can be solved for the remaining m variables. These variables we call 'basic', those which we have set equal to 0 are called 'non-basic'. The essence of the method conists in finding at each step 'feasible', that is non-negative, variables. At each step, one of the basic variables is made non-basic, and vice versa: we 'exchange' two variables.

Let the problem to be solved be

maximize $B = b_1x_1 + \ldots + b_nx_n + b_0$

subject to

$$a_{11}x_1 + \ldots + a_{1n}x_n + x_{n+1} = c_1$$

$$- - - - - - - - -$$

$$a_{m1}x_1+\ldots+a_{mn}x_n+x_{n+m}=c_m$$

$$x_i \geq 0 \ (i=1,2,\ldots,n,n+1,\ldots,n+m)$$

The variables $x_{n+1},\ldots, x_{n+m}$ are slack variables.

To begin with, we make $x_1,\ldots, x_n$ non-basic and $x_{n+1},\ldots, x_{n+m}$ basic. Then

$$x_i=0 \ (i=1,\ldots,n) \text{ and } x_{n+j}=c_j \ (j=1,2,\ldots,m).$$

We assume that the c_j are positive.

We wish to maximize B and to find those basic variables which achieve this. The non-basic variables are zero, and no variable may have a lower value. If one of the b_i is positive, say $b_{i_0}>0$, the it is useful to increase x_{i_0}.

How far can we increase x_{i_0}? There is a limit to be considered, because no other variable must become negative as a consequence.

Assume that x_{i_0} has been increased to a positive value, while all other non-basic variables remain at zero. It follows that

$$x_{n+j}=c_j-a_{ji_0}x_{i_0} \text{ and } B=b_{i_0}x_{i_0}+b_0.$$

Scrutinizing x_{n+j} $(j=1,\ldots, m)$ we find that those in which a_{ji_0} is negative, have increased; there is no risk of their value becoming negative. But where a_{ji_0} is positive, then $x_{i_0}>c_j/a_{ji_0}$ means $x_{n+j}<0$. Therefore x_{i_0} may not be made larger than the smallest of c_j/a_{ji_0} for those j, for which $a_{ji_0}>0$. If this value is obtained for $j=j_0$, then increasing x_{n+j_0} to its allowed limit implies making x_{n+j_0} zero. We exchange x_{i_0} and x_{n+j_0}.

For simplicity of notation, let $i_0=1$ and $j_0=1$. We express once more the basic variables in terms of the non-basic ones, and obtain

$$x_1+\frac{1}{a_{11}}x_{n+1}+\frac{a_{12}}{a_{11}}x_2+\ldots+\frac{a_{1n}}{a_{11}}x_n=\frac{c_1}{a_{11}}$$

$$x_{n+j}-\frac{a_{j1}}{a_{11}}x_{n+1}+\left(a_{j2}-\frac{a_{12}a_{j1}}{a_{11}}\right)x_2+\ldots+$$

$$\left(a_{jn}-\frac{a_{1n}a_{j1}}{a_{11}}\right)x_n=c_j-\frac{a_{j1}}{a_{11}}c_1 \ (j=2,\ldots,m) \qquad (*)$$

$$B=\frac{b_1}{a_{11}}x_{n+1}-\left(b_2-\frac{a_{12}b_1}{a_{11}}\right)x_2-\ldots-\left(b_n-\frac{a_{1n}b_1}{a_{11}}\right)x_n$$

$$=b_0+\frac{b_1c_1}{a_{11}} \qquad (**)$$

This finishes one step.

We look again at the coefficients of the non-basic variables in the expression for B, and, using the same criteria as before, proceed if necessary.

It can be proved that if an optimal solution exists, then at least one basic optimal solution exists as well. So we do not lose anything by restricting ourselves to basic solutions.

If a coefficient of B is zero, then an increase of the corresponding non-basic variable is possible, but the value of B will not change. Thus there may be more than one optimal basic solution. In such a case a convex combination of the optimal basic solutions will also be optimal.

Example 1
maximize $x_0 = x_1 + x_2 + x_3$
subject to

$$x_4 + x_1 - x_2 + 3x_3 = 1$$

$$x_5 + 3x_1 + 5x_2 - 3x_3 = 1$$

$$x_6 + 6x_1 + 2x_2 - 2x_3 = 1$$

all $x_i \geq 0$.

We write this in tableau form, rows for basic and columns for non-basic variables.

	x_1	x_2	x_3	
x_4	1	−1	3	1
x_5	3	5	−3	1
x_6	6*	2	−2	1
x_0	−1	−1	−1	0

The exchange of variables, and the transformation of the tableau can be described as follows:

Choose a column with a negative value in the x_0 row (for instace, in the x_1 column).

Find the smallest ratio dividing the value in the last column by the positive value of the column chosen in the same row. (For instance, min (1/1, 1/3, 1/6) = 1/6.)

The variable to be exchanged is that of the row producing that minimum. (In the example, exchange x_6 for x_1.) The value in the intersection of the respective row and column is the 'pivot', marked by an asterisk.

The rules of transformation, described by formulae (*) and (**), are these:

(1) Replace the pivot by its reciprocal.
(2) Divide all other entries in the pivotal row by the pivot.
(3) Divide all other entries in the pivotal column by the pivot and change the sign.
(4) From the remaining values, a_{ij} say, subtract $(a_{ij_0} a_{i_0 j})/a_{i_0 j_0}$.

$a_{i_0 j_0}$ is, of course, the pivot.

In our example, we obtain the following sequence of tableaux:

	x_6	x_2	x_3	
x_4	−1/6	−4/3	10/3	5/6
x_5	−1/3	4*	−2	1/2
x_1	1/6	1/3	−1/3	1/6
x_0	1/6	−2/3	−4/3	1/6

	x_6	x_5	x_3	
x_4	−1/3	1/3	8/3*	1
x_2	−1/8	1/4	−1/2	1/8
x_1	5/24	−1/12	−1/6	1/8
x_0	1/12	1/6	−5/3	1/4

	x_6	x_5	x_4	
x_3	−1/8	1/8	3/8	3/8
x_2	−3/16	5/16	3/16	5/16
x_1	3/16*	−1/16	1/16	3/16
x_0	−1/8	3/8	5/8	7/8

	x_1	x_5	x_4	
x_3	2/3	1/12	5/12	1/2
x_2	1	1/4	1/4	1/2
x_6	16/3	−1/3	1/3	1
x_0	2/3	1/3	2/3	1

This is the final tableau: all values in the x_0 row are positive. The answer is

$$x_1 = 0 \text{ (non-basic)}, x_2 = 1/2, x_3 = 1/2, x_4 = 0, x_5 = 0, x_6 = 1. \text{ Maximum of } x_0 = 1.$$

It turns out, that x_1, which we have made basic at the first step, has become non-basic again, and we could have chosen some other variable to begin with. But of course, we could not know this.

Example 2
Maximize $x_0 = x_1 + x_2 + x_3 + x_4$.
subject to

$$x_1 + 3x_2 - 2x_3 + x_4 \leq 1$$

$$-x_1 + x_2 + x_3 + 4x_4 \leq 1$$

$$4x_1 + x_2 + x_3 - 3x_4 \leq 1$$

$$x_1 - 2x_2 + 5x_3 + x_4 \leq 1$$

all $x_i \geq 0$.

We introduce slack variables x_5, x_6, x_7, x_8. These are the basic variables to begin with, their initial values are 1.

The final tableau reads

	x_7	x_5	x_6	x_4	
x_2	1/25	1/5	9/25	38/25	3/5
x_3	4/25	−1/5	11/25	27/25	2/5
x_1	1/5	0	−1/5	−7/5	0
x_8	−23/25	7/5	−32/25	1/25	1/5
x_0	2/5	0	3/5	1/5	1 .

The answer is

$$x_1 = 0, x_2 = 3/5, x_3 = 2/5, x_8 = 1/5.$$

However, this is not the whole story. There is a value 0 in the x_0 row. Therefore, x_5 may be increased. We can exchange x_5 for x_8, and obtain

$$x_2 = 4/7, x_3 = 3/7, x_1 = 0, x_5 = 1/7.$$

The value of x_0 is, of course, unchanged. Any values $(x_1, x_2, x_3, x_4) = (0, v, 1-v, 0)$ with $4/7 \le v \le 3/5$ are optimal.

As a counterpart to maximizing $b_1x_1 + \ldots + b_nx_n$
subject to

$$a_{11}x_1 + \ldots + a_{1n}x_n \le c_1$$

$$- - - -$$

$$a_{m1}x_1 + \ldots + a_{mn}x_n \le c_m$$

all $x_i \ge 0$
consider its 'dual' problem
minimize $c_1 y_{n+1} + \ldots + c_m y_{n+m}$
subject to

$$a_{11}y_{n+1} + \ldots + a_{m1} y_{n+m} \ge b_1$$

$$a_{1n} y_{n+1} + \ldots + a_{mn} y_{n+m} \ge b_n$$

all $y_{n+j} \ge 0$.

It is, of course, immaterial which subscripts we choose, and for reasons which will emerge, we start with the subscripts of y following those of x.

Example 3
This is the dual to Example 1.
Minimize $y_4 + y_5 + y_6 = y_0$
subject to

$$y_4 + 3y_5 + 6y_6 \ge -1$$

$$-y_4 + 5y_5 + 2y_6 \geq -1$$

$$3y_4 - 3y_5 - 2y_6 \geq -1$$

$$y_4, y_5, y_6 \geq 0.$$

Introduce slack variables y_1, y_2, and y_3, and let them be the first non-zero variables. We obtain the tableau

	y_4	y_5	y_6	
y_1	−1	−3	−6	−1
y_2	1	−5	−2	−1
y_3	−3	3	2	−1
y_0	−1	−1	−1	0

The variables y_1, y_2, and y_3 have negative values. On the other hand, all values in the y_0 row are negative: just the sort of entries we want to have in the final tableau, when we minimize.

Let us compare the first tableaux of examples 1 and 3. The rows of the x-tableau equal the columns of the y-tableau with changed signs, except that the x_0-row is precisely the last y-column. The present values of x_0 and y_0 are equal, viz. 0. We call such two tableaux dual to one another.

Now write down the dual to the second tableau in example 1.

	y_4	y_5	y_1	
y_6	1/6	1/2	−1/6	1/6
y_2	4/3	−4	−1/3	−2/3
y_3	−10/3	2	1/3	−4/3
y_0	−5/6	−1/2	−1/6	1/6

This tableau is also the result of the first y-tableau by the rules of the simplex method, if we choose the pivot, −6, in the y_1 row and y_6 column, corresponding to the pivot 6 in the first x-tableau.

Since the rules of the simplex method are algebraic rules of substitution, unaffected by the signs of the elements on which they work, the lines of the last tableau are again transformations of the constraints.

Without reference to the x-tableau, we can describe the generation of the second tableau from the first as follows:

(a) choose the row of a variable with negative value;
(b) the pivot must be negative
(c) choose, for the exchange of variables, a column in which the ratio of the value in the y_0 row, divided by that in the row to be exchanged is smallest;
(d) apply the rules of the simplex transformation.

Carrying on with this ‘dual simplex method’ we find that at each step the resulting tableau is dual to that of the x-tableau at the same step. Eventually, we obtain

	y_3	y_2	y_6	
y_1	−2/3	−1	−16/3	2/3
y_5	−1/12	−1/4	1/3	1/3
y_4	−5/12	−1/4	−1/3	2/3
y_0	−1/2	−1/2	−1	1

All values of the variables are now non-negative, and all values in the y_0 row are negative. This is, therefore, the final tableau.

$$y_1 = 2/3,\ y_2 = 0,\ y_3 = 0,\ y_4 = 2/3,\ y_5 = 1/3,\ y_6 = 0.$$

The minimum of y_0 equals the maximum of x_0.

Because of the relationship between the dual tableaux, the values of the y-variables equal those in the x_0 row, in the columns with the same subscripts of x and y. Therefore, if we are interested only in the final values of the variables, it is unnecessary to work out both final tableaux; the answer can be found for the dual problem, by looking at the x_0-row in the final tableau of the x-problem, the primal problem.

We discuss a few more aspects of looking at the dual of Example 2.

Example 4
Minimize $y_0 = y_5 + y_6 + y_7 + y_8$
subject to

$$y_5 - y_6 + 4y_7 + y_8 \geq 1$$

$$3y_5 + y_6 + y_7 - 2y_8 \geq 1$$

$$-2y_5 + y_6 + y_7 + 5y_8 \geq 1$$

$$y_5 + 4y_6 - 3y_7 + y_8 \geq 1$$

We introduce slack variables y_1, y_2, y_3, y_4 and obtain the tableau

	y_5	y_6	y_7	y_8	
y_1	−1	1	−4	−1	−1
y_2	−3	−1	−1	2	−1
y_3	2	−1	−1	−5	−1
y_4	−1	−4	3	−1	−1
y_0	−1	−1	−1	−1	0

Computed by the dual simplex method, the final tableau will be

	y_2	y_3	y_1	y_8	
y_7	–1/25	–4/25	–1/5	23/25	2/5
y_5	–1/5	1/5	0	–7/5	0
y_6	–9/25	–11/25	1/5	32/25	3/5
y_4	–38/25	–27/25	7/5	–1/25	1/5
y_0	–3/5	–2/5	0	–1/5	1

The values of the basic variables are

$$y_7 = 2/5, y_5 = 0, y_6 = 3/5, y_4 = 1/5$$

and the minimum of the objective function is 1.

We could have found this out by looking at the final tableau of example 2, the dual of example 4.

Example 2 had an alternative basic solution, and so has example 4.

If we use 0, in the y_0 row and y_1 column to transform the tableau once more, using 7/5 as the pivot, we obtain

$$y_7 = 3/7, y_5 = 0, y_6 = 4/7, y_4 = 1/7$$

Similarly, we can get an alternative y_0 row in example 4, by making use of $y_5 = 0$ to continue with the dual simplex method. The y_0 will then become

$$-4/7 \quad -3/7 \quad 0 \quad -1/7 \quad 1.$$

Combining all these results, we have four pairs of optimal basic variables (and, of course, all their convex combinations).

(1) $x_1 = 0, x_2 = 3/5, x_3 = 2/5, x_8 = 1/5$
with $y_4 = 1/5, y_5 = 0, y_6 = 3/5, y_7 = 2/5$

(2) $x_1 = 0, x_2 = 4/7, x_3 = 3/7, x_5 = 1/7$
with $y_4 = 1/5, y_5 = 0, y_6 = 3/5, y_7 = 2/5$

(3) $x_1 = 0, x_2 = 3/5, x_3 = 2/5, x_8 = 1/5$
with $y_4 = 1/7, y_5 = 0, y_6 = 4/7, y_7 = 3/7$

(4) $x_1 = 0, x_2 = 4/7, x_3 = 3/7, x_5 = 1/7$
with $y_4 = 1/7\ y_5 = 0, y_6 = 4/7, y_7 = 3/7.$

Appendix II: Permutation groups

A group is a set of elements with a law of composition, which we call multiplication, such that

—for every ordered pair of elements a,b there exists a unique element c of the group, the 'product', written $c = a \cdot b$;
—the associative law $(ab)c = a(bc)$ holds for all elements of the group
—there exists an element of the group, the 'unit', written 1, such that $a \cdot 1 = 1 \cdot a = a$ for every element of the group; and
—there exists an element to every element a, written a^{-1}, its 'inverse', such that $a \cdot a^{-1} = a^{-1} \cdot a = 1$.

If, furthermore, $a \cdot b = b \cdot a$ holds for any pair of elements a and b, the group is called 'cummutative'.

The number of elements of a group is called its 'order'.

The n! permutations of n objects form such a group. If we denote the objects by 1,2 ,..., n, then the permutations, the elements of the permutation group, may be written

$$\begin{pmatrix} 1 & 2 & \dots & n \\ a_1 & a_2 & \dots & a_n \end{pmatrix}$$

which means, that i is changed into the element a_i. The set of all such permutations is the 'symmetric group' of n objects.

A permutation of the form

$$\begin{pmatrix} 1 & 2 & \dots & m-1 & m \\ 2 & 3 & \dots & m & 1 \end{pmatrix}$$

is called 'cyclic' of 'degree' m. The elements not quoted remain unaltered. Such a cyclic permutation may also be written

$$(1,2,\dots,m-1,m) = (2,3,\dots,m,1) = \dots = (m,1,2,\dots,m-1).$$

A cycle of degree 2 is a 'transposition', and a cycle of degree m is the product of $(m-1)$ transpositions

$$(a_1,a_2,\ldots,a_m)=(a_1,a_2)\cdot(a_1,a_3)\ \ldots\ (a_1,a_m).$$

Any permutation, cyclic or not, is the product of disjoint cycles. For instance,

$$\begin{pmatrix}1\,2\,3\,4\,5\\2\,1\,4\,5\,3\end{pmatrix}=(1\,2)\cdot(3\,4\,5).$$

Consider n objects $x_1,\ldots,x_n$ and their product of differences

$$D=(x_1-x_2)\,(x_1-x_3)\ \ldots\ (x_1-x_n)\,(x_2-x_3)\ \ldots\ (x_2-x_n)\ \ldots\ (x_{n-1}-x_n).$$

A permutation of the objects either leaves D unchanged or multiplies it by -1. in the former case the permutation is called 'even', in the latter case it is called 'odd'. A product of two even or of two odd permutations is even, and that of an even and an odd permutation is odd. Hence a cycle of degree m has odd parity if m is even, and even parity if m is odd. Also, every even permutation is the product of cycles of degree 3, each of which is even.

We prove the last statement:

Every permutation is the product of disjoint cycles, and every cycle a product of transpositions; an even permutation is the product of an even number of transpositions. Take the transpositions in pairs, and let one pair be say $(u_1,v_1)\,(u_2,v_2)$.

This equals $(u_1,v_1)\,(v_1,u_2)\,(u_2,v_1)\,(u_2,v_2)$, because the product of the second and the third transposition is unity.

Combining the first two transpositions into a cycle of degree 3, we have (u_1,u_2,v_1), and combining the last two transpositions into a cycle we have (u_2,v_1,v_2). Thus, from the product of two transpositions, we obtain a product of two cycles of degree 3. Q.E.D.

Appendix III: Fibonacci numbers

Fibonacci numbers are defined by the recurrence

$$F_0 = 0, F_1 = 1, F_{n+2} = F_{n+1} + F_n \ (n = 0, 1, ..).$$

It follows that $F_{n+1} < 2F_n$ when $n > 3$.

The sequence starts with

$$0,\ 1,\ 1,\ 2,\ 3,\ 5,\ 8,\ 13,\ 21,\ 34,\ 55,\ \ldots$$

A great number of relationships betwee Fibonacci numbers are known (see, for instance, Vajda, 1989). We mention here those which are relevant in the game of Fibonacci Nim.

Zeckendorf's Theorem
Any positive integer N can be expressed 'canonically' as a sum of Fibonacci numbers

$$N = F_{k_1} + F_{k_2} + \ldots + F_{k_r}$$

where $k_{i+1} \le k_i - 2$ $(i = 1, 2, ..., r-1)$, (i) and $k_r \ge 2$. (ii)
Such a representation is unique.

Proof: If N is a Fibonacci number, the theorem is trivial. It is found, by inspection, to be true for $N = 3 + 1 = F_4 + F_2$. Assume it to be true for all integers up to and including F_n, and let $F_{n+1} \ge N > F_n$.

Now $N = F_n + (N - F_n)$, and $N \le F_{n+1} < 2F_n$, that is $N - F_n < F_n$.

So $N - F_n$ can be expressed in the form

$$F_{t_1} + F_{t_2} + \ldots + F_{t_r},\ t_{i+1} \le t_i - 2, t_r \ge 2,$$

and $N = F_n + F_{t_1} + \ldots + F_{t_r}$.

We can be certain that $n > t_1 + 2$, because if we had $n = t_1 + 1$, then $F_n + F_{t_1+1} = 2F_n$. But this is larger than N.

In fact, F_n must appear in the canonical representation of N, because no sum of smaller Fibonacci numbers, obeying (i) and (ii), could add up to N. This follows, if n is even, say 2k, from

$$F_{2k-1} + F_{2k-3} + \ldots + F_3 = (F_{2k} - F_{2k-2}) + (F_{2k-2} - F_{2k-4}) + \ldots + (F_4 - F_2) = F_{2k} - 1;$$

and if n is odd, say $2k - 1$, it follows from

$$F_{2k} + F_{2k-2} + \ldots + F_2 = (F_{2k+1} - F_{2k-1}) + \ldots + (F_3 - F_1) = F_{2k+1} - 1$$

Again, in the representation of $N - F_n$, the largest F_i not exceeding $N - F_n$ must appear, and it cannot be F_{n-1}. This proves uniqueness by induction.

We prove now two properties of the canonical representation.

Property I

If $F_{k_i} > F_{k_j}$, then $F_{k_i} > 2F_{k_j}$.

Proof

$k_j \leq k_i - 2$ by (i), and $F_{k_i} = F_{k_i-1} + F_{k_i-2} > 2F_{k_i-2}$.

Property II

Let $F_{n+1} - N = D$. Then $2D > F_{k_r}$.

Proof

$$D \geq F_{k_1+1} - F_{k_1} - F_{k_i-2} - F_{k_i-4} - \ldots - F_{k_r} =$$

$$F_{k_1-1} - F_{k_1-2} - \ldots - F_{k_r} =$$

$$F_{k_1-3} - \ldots - F_{k_1-2j} - F_{k_r} =$$

$$F_{k_1-2j-1} - F_{k_r} \geq F_{k_1-2j-3},$$

hence $2D > 2F_{k_1-2k_j-3} > F_{k_1-2j-2} > F_{k_r}$.

Example

$N = 19 = 13 + 5 + 1 = F_7 + F_5 + F_2$.

(I) $13 > 2 \times 5, 5 > 2 \times 1$.

(II) $D = 21 - 19 = 2, 2D = 4 > 2 \times 1$.

Appendix IV: A little graph theory

A graph is a set of vertices and edges (arcs), the latter joining some pairs of vertices. In 'directed' graphs, the edges have a direction from a vertex a to a vertex b, and b is the 'follower' of a.

A graph is called finite if the number of vertices is finite, and a vertex is follower-finite if it has only a finite number of followers. A progressively finite graph is one in which all paths starting from any vertex are finite, that is terminating in a vertex without a follower, and no path contains a loop. A graph without loops or multiple edges between two vertices is called simple. The relationship between progressively finite graphs and finite games needs no comment.

A tree is a graph with one vertex, its root, defined, with a unique path from the root to any other vertex; a tree with n vertices contains $n - 1$ arcs. A graph consisting of a number of trees is called a forest.

The kernel of a graph is a set of vertices such that there is no edge between any two vertices of the set, and there is an arc from any vertex outside the set to a vertex within the set. Every progressively finite graph has a kernel (Tucker, 1980, pp. 364–365). If the vertices of a graph represent positions in a game, and the directed arcs represent possible moves, then the kernel is the set of winning positions.

The Grundy function $g(x)$ of the vertex x is a function such that $g(x)$ is the smallest non-negative integer not in the set $g(y)$ of the followers y of x. A vertex without a follower has $g(x) = 0$. A progressively finite graph has a Grundy function, which can be computed inductively starting from a terminal position. If a vertex is follower-finite, then its $g(x)$ cannot exceed the number of followers of x.

Another area of graph theory deals with combinatorial properties. Let the vertices of a graph represent persons. Colour an edge green, if it connects two individuals who know each other, and red, if they do not. Every one of the edges of a complete graph will be coloured, either green or red. A pretty theorem states that if in a complete graph of six vertices all fifteen edges are coloured, then at least one triangle is such that all three sides have the same colour. A simple proof runs as follows:

From a vertex a there start five edges. At least three of them must have the same colour, say red. Let them be those which connect a with b, c, and d. If one of the edges *bc, bd, cd* is red, perhaps *bc*, then *abc* is a red triangle. But if all these edges are green, then *bcd* is a green triangle. (Bostwick, 1958).

If there are six people at a party, some pair might already know one another, and some pair might not. The theorem which we have proved tells us that there must be at least one triple who all know each other, or at least one triple who are mutually unknown to each other. This application of the theorem explains the title of the papers by Goodman (1959) and by Harary (1972).

Goodman (1959) has proved that in a completely coloured graph with N vertices and two colours there must be at least

$n(n-1)(n-2)/3$ coloured triangles when $N = 2n$
$2n(n-1)(4n+1)$ coloured triangles when $N = 4n+1$
$2n(n+1)(4n-1)$ coloured triangles when $N = 4n+3$

The first of these three cases implies two triangles when $N = 6$.

By a detailed study of all possible patterns Harary (1972) has shown for $N = 6$ that there exist two-colourings with exactly two monochromatic triangles with 0, 1, or 2 common points. The two triangles have different colours if and only if they have just one single common point. (When they have two common points, and hence a common line, then they must obviously have the same colour.)

Appendix V: Termination games are equivalent to Nim (Holladay, 1957)

We accept, without proving it here, that in a termination game all positions are either safe, or unsafe, in the following sense:

(a) any move from a safe position leads to an unsafe position;
(b) from any unsafe position there exists a move which leads to a safe position.
(c) Every terminating position is safe.

Lemma.
Given a position p in a finite game G, we define i to be the value of p if in the disjunctive sum of the finite game G and one-pile Nim, (p, i) is a safe position. Given p, the integer i exists and is unique.

Proof.
(a) Uniqueness. If there were two such sizes in one-pile Nim, say i and $j > i$, then one could move from a safe position (p, i) to another safe position (p, j), which would contradict the property of a safe position.
(b) Existence. If for some p there were no i such that (p, i) is safe, then there would exist a position (p', i) which is safe, owing to the property of an unsafe position. This would be a different p' for different i, because the same p' cannot have two different values, as we have just shown. But the set of possible p' is finite, while that of i is infinite. Hence there must exist, for any p, an i such that (p, i) is safe.

Now consider positions in a compound game, with components $G_1, \dots, G_k$. We show that if $(p_1, \dots, p_k)$ is a position in the disjunctive sum of these games, and if i_j is the value of p_j $(j = 1, \dots, k)$, then $(p_1, \dots, p_k)$, is safe in the compound game, when $(i_1, \dots, i_k)$ is safe in k-pile Nim.

Proof. To begin with, let $k = 2$. G is the disjunctive sum of G_1 and G_2. Let p_1 be safe in G_1, and p_2 unsafe in G_2. Then (p_1,p_2) cannot be safe in G. If p_2 is changed to be safe in G_2, then the next move of the opponent will leave one of the components, positions in G_1 and G_2, unsafe. It can again be made safe, and so on, until eventually p_1 as well as p_2 are safe in the final position.

Returning to the disjunctive sum of G_1 ,..., G_k, $k > 2$, we can find for each p_j an i_j such that (p_j,i_j) is safe, and by repeating the above argument for $k = 2$, $(p_1,i_1$,..., $p_k,i_k)$ is safe in the disjunctive sum of G_1 ,..., G_k and k times one-pile Nim. Hence $(p_1$,..., $p_k)$ is safe if and only if $(i_1$,..., $i_k)$ is safe, that is winning, in k-pile Nim.

References

Alexander, C. (1990). Non-cooperative finite games, Ch. 10 of *Handbook of applicable mathematics. Supplement.* Wiley.

Alfred, Br. U. (1963) Fibonacci Nim. *The Fibonacci Quarterly* **1**, 63.

Ali, I., Cook, W. D. and Kress, M. (1986). On the minimum violations of ranking of a tournament. *Management Science* **32**, 660–672.

Beasley, J. D. (1962). Some notes on Solitaire. *Eureka* **25** 13–18.

Beasley, J. D. (1985). *The ins and outs of Peg Solitaire.* Oxford University Press.

Beasley, J. D. (1989). *The Mathematics of Games.* Oxford University Press.

Beck, A., Bleicher, M. and Crow, D. (1969). *Excursions into mathematics,* pp. 327–329. Worth.

Bergholt, E. (1920). *The game of Solitaire.* Routledge.

Berlekamp, E. R., Conway, J. H. and Guy, R. K. (1982). *Winning ways for your mathematical plays.* Academic Press.

Blackwell, D. and Girshick, M. A. (1954). *Theory of games and statistical decisions.* Wiley.

Bostwick, C. W. (1958). Problem E 138. *Amer. Math. Mthly* **65**. Solved by J. Rainswater **66** (1959) 14.

Bouton, C. L. (1902). Nim, a game with a complete mathematical theory. *Annals of Math.* **3**, 35–39.

Connell, I. G. (1959). A generalization of Wythoff's game. *Can. Math. Bull.* **2**, 181–190.

Conway, J. H. (1976). On numbers and games. Academic Press.

Csirmaz, L. (1980). On a combinatorial game with an application to Go-Moku. *Discr. Math.* **29**, 19–23.

Dawson, T. R. (1934). Problem 1603. *Fairy Chess Review.* December, p. 94.

deBruijn, N. G. (1972). A solitaire game and its relation to a finite field. *J. Recr. math.* **5**, 133–137.

de Carteblance, F. (1970). The Princess and the Roses. *J. Recr. Math.* **3**, 238–239.

de Carteblance, F. (1974). The Roses and the Princess. *J. Recr. Math.* **7**, 295–298.

Descartes, B. (1953). Why are series musical? *Eureka* **16**, 18–21.

Dudeney, H. E. (1907). *The Canterbury Puzzles and other curious problems.* Nelson.

Dudeney, H. E. (1917). *Amusements in mathematics*. Nelson.

Duvdevani, N. and Fraenkel, A. S. (1989). Properties of *k*-Welter's game. *Discr. Math.* **76,** 197–221.

Erdös, P. and Selfridge, J. L. (1973). On a combinatorial game. *J. Comb. Theory (B)* **14,** 298–301.

Evans, R. (1974). A winning opening in reverse Hex. *J. Recr. Math.* **7**(3), 189–192.

Evans, R. (1975–6). Some variants of Hex. *J. Recr. Math.* **8**(2), 120–122.

Farkas, J. (1901–2). Theorie der einfachen Ungleichungen. *J. reine und angewandte Math.* **124,** 1–27.

Ferguson, T. S. (1974). On sums of graph games with last player losing. *Int. J. of Game Theory*. **3,** 159–167.

Fraenkel, A. S. and Borosh, I. (1973). A generalization of Wythoff's game. *J. Comb. Theory. (A)* **15,** 175–191.

Fraenkel, A. S., Loebl, M. and Nešetřil, J. (1988). Epidemiography II. *J. Comb. Theory (A)* **49**(1), 129–144.

Fraenkel, A. S. and Lorberbom, M. (1989). Epidemiography with various growth functions. *Discr. Appl. Math.* **25,** 53–71.

Fraenkel, A. S. and Nešetřil, J. (1985). Epidemiography. *Pac. J. math.* **118**(2), 369–381.

Gale, D. (1974). A curious Nim-type game. *Amer. Math. Mthly* **81,** 876–879.

Gale, D. (1979). The game of Hex ad the Brouwer fixed point theorem. *Amer. Math. Mthly* **86,** 818–827.

Gardner, M. (1959). *The Scientific American book of mathematical puzzles and diversions*. Simon and Schuster.

Gardner, M. (1966). *New mathematical diversions from the Scientific American* pp. 212–216.

Gardner, M. (1973). Mathematical games. *Scientific American* **228,** Jan. 110–111, Febr. 109.

Gardner, M. (1974). Mathematical games. *Scientific American* **230.** Aug. 106–108.

Goddard, S. T. (1983). Ranking tournaments and group decision making. *Management Science* **29,** 1384–1392.

Goodman, A. W. (1959). On sets of acquaintances and strangers at any party. *Am. Math. Mthly* **66,** 778–783.

Grundy, P. M. (1939). Mathematics and Games. *Eureka* **2,** 9–11.

Grundy, P. M. and Smith, C. A. B. (1956). Disjunctive games with the last player losing. *Proc. Cambridge Phil. Soc.* **52,** 527–533.

Guy, R. K. (1976). Twenty questions concerning Conway's Sylver Coinage. *Amer. Math. Mthly* **83,** 634–637.

Guy, R. K. and Smith, C. A. B. (1956). The g-values of various games. *Proc. Cambridge Phil. Soc.* **52,** 514–526.

Hales, A. W. and Jewett, R. I. (1963). Regularity and positional games. *Trans. Amer. Math. Soc.* **106,** 222–229.

Hall, P. (1935). On representations of subsets. *J. London Math. Soc.* **10,** 26–30.

Harary, F. (1972). The two-triagular case of the acquaintance graph. *Math. Magazine* **45**(3), 130–135.

Hentzel, I. R. (1973) Triangular Puzzle Peg. *J. Recr. Math.* **6,** 280–283.

Holladay, J. C. (1957). Cartesian products of termination games. *Annals of Math. Studies* **39,** 189–200.

Jenkyns, T. A. and Mayberry, J. P. (1980). The skeletion of an impartial game and the Nim-function of Moore's Nim. *Intern. J. of Game Theory* **9,** 51–63.

Kahane, J. and Fraenkel, A. S. (1987). A generalization of Welter's game. *J. Comb. Theory (A)* **46,** 1–20.

Kenyon, J. C. (1967). *Nimlike games and the Sprague–Grundy theory.* Thesis, University of Calgary, Alberta.

Kowalewski, G. (1930). *Alte und neue mathematische Spiele.* Leipzig.

Landau, H. G. (1953). On dominance relations and the structure of animal societies: III. The conditions for a score sequence. *Bull. Math. Biophysics* **15,** 114–118.

Lasker, E. (1931). *Brettspiele der Völker.* Berlin.

Lehman, A. (1964). A solution of the Shannon Switching Game. *J. Soc. Ind. Appl. Math.* **12,** 687–725.

Liebeck, H. (1971). Some generalizations of the 14–15 puzzle. *Math. Magazine* **44,** 185–189.

Lucas, E. (1960). *Récréations mathématiques.* 2nd. ed. Blanchard, Paris.

Moon, J. W. (1963). An extension of Landau's theorem on tournaments. *Pac. J. Math.* **13,** 1343–1345.

Moore, E. H. (1910). A generalization of the game called Nim. *Ann. of Math.* (2) **11,** 93–94.

O'Beirne, T. H. (1965). *Puzzles and paradoxes.* Chapter 9. Oxford University Press.

Pelletier, D. H. (1987) Merlin's Magic Square. *Amer. Math. Mthly* **94,** 143–150.

Reiss, M. (1857). Beiträge zur Theorie des Solitär-Spiels. *Crelle's J.* **54,** 344–379.

Rounds, E. M. and Yau, S. S. (1974). A winning strategy for Sim. *J. Recr. Math.* **7**(5), 193–202.

Schuh, F. (1952). The game of divisions. *Nieuw Tidscrift voor Wijskunde* **39,** 299–304.

Schuh, F. (1968). *The master-book of mathematical recreations.* Transl. F. Göbel. Dover.

Schwenk, A. J. (1970). Take-away games. *The Fibonacci Quart.* **8,** 225–234.

Simmons, G. J. (1969). The game of Sim. *J. Recr. Math.* **2**(2), 66.

Smith, C. A. B. (1966). Graphs and composite games. *J. Comb. Theory* **1,** 51–81.

Smith, C. A. B. (1968). Compound games with counters. *J. Recr. Math.* **1,** 67–77.

Spitznagel, E. L. (1967). A new look at the Fifteen puzzle. *Math. Magazine* **40,** 171–173.

Sprague, R. (1935–1936) Über mathematische Kampfspiele. *Tohoku Math. J.* **41,** 438–444.

Sprague, R. (1937). Über zwei Abarten von Nim. *Tohoku Math. J.* **43,** 351–359.

Sprague, R. (1948–1949). Bemerkungen über eine spezielle Abel'sche Gruppe. *Math. Zeitschrift* **57,** 82–84.

Sprague, R. (1963) *Recreations in mathematics. (Unterhaltsame Mathematik),* tranl. O'Beirne. Vieweg.

Suttner, K. (1980). Linear cellular automata and the Garden of Eden. *Math. Intelligencer* **11,** 49–53.

Sylvester, J. J. (1884). Mathematical Questions. *Educational Times* **41,** 21.

Tucker, A. (1980). *Applied Combinatorics,* pp. 355–359. Wiley.

Úlehla, J. (1980). A complete analysis of von Neumann's Hackendot. *Intern. J. of Game Theory* **9,** 107–113.

Vajda, S. (1981). *Linear Programming.* Chapman and Hall.

Vajda, S. (1989). *Fibonacci & Lucas numbers, and the Golden Section.* Ellis Horwood, Chichester.

von Neumann, J. (1928). Zur Theorie der Gesellschaftsspiele. *Math. Annalen* **100,** 295–320.

von Neumann, J. and Morgenstern, O. (1953). *The theory of games and economic behavior* (2nd ed) Princeton University Press.

Wei, T. H. (1952). *The algebraic foundations of ranking theory.* Thesis, Cambridge University.

Welter, C. P. (1952). The advancing operation in a special Abelian group. *Nederland Acad. Wetensc. Proc. (A),* **55,** (*Indag. Mat.* **14,** 304–314).

Welter, C. P. (1954). The theory of a class of games on a sequence of squares in terms of the advancing operation in a special group. *Nederl. Acad. Wetensc. Proc. (A),* **57** (*Indag. Mat.* **16,** 194–200).

Whinihan, M. J. (1963) Fibonacci Nim. *The Fibonacci Quarterly* **1,** 9–13.

Wiegleb, J. C. (1779). *Anhang von dreyen Solitärspielen. Unterricht in der natürlichen Magie.* J. N. Martius. Berlin.

Wilkinson, J. H. (1965). *Algebraic eigenvalue problems.* Clarendon Press, Oxford.

Wilson, N. Y. (1959). A number problem. *Proc. Edinburgh Math. Soc.* **II**(4), 11–14.

Wilson, R. M. (1974). Graph puzzles, homotopy, and the alternating group. *J. Comb. Th. (B)* **16,** 86–96.

Wythoff, M. A. (1907). A modification of the game of Nim. *Nieuw. Arch. Wijsk.* **7,** 199–202.

Zetters, T. G. L. (1980). 8 (or more) in a row. Problem 510. *Amer. Math. Mthly* **87,** 575–576.

Dictionary of games

Index

A CATALOG OF SELECTED
DOVER BOOKS
IN SCIENCE AND MATHEMATICS

Mathematics

FUNCTIONAL ANALYSIS (Second Corrected Edition), George Bachman and Lawrence Narici. Excellent treatment of subject geared toward students with background in linear algebra, advanced calculus, physics and engineering. Text covers introduction to inner-product spaces, normed, metric spaces, and topological spaces; complete orthonormal sets, the Hahn-Banach Theorem and its consequences, and many other related subjects. 1966 ed. 544pp. 6⅛ x 9¼. 0-486-40251-7

ASYMPTOTIC EXPANSIONS OF INTEGRALS, Norman Bleistein & Richard A. Handelsman. Best introduction to important field with applications in a variety of scientific disciplines. New preface. Problems. Diagrams. Tables. Bibliography. Index. 448pp. 5⅜ x 8½. 0-486-65082-0

VECTOR AND TENSOR ANALYSIS WITH APPLICATIONS, A. I. Borisenko and I. E. Tarapov. Concise introduction. Worked-out problems, solutions, exercises. 257pp. 5⅝ x 8¼. 0-486-63833-2

AN INTRODUCTION TO ORDINARY DIFFERENTIAL EQUATIONS, Earl A. Coddington. A thorough and systematic first course in elementary differential equations for undergraduates in mathematics and science, with many exercises and problems (with answers). Index. 304pp. 5⅜ x 8½. 0-486-65942-9

FOURIER SERIES AND ORTHOGONAL FUNCTIONS, Harry F. Davis. An incisive text combining theory and practical example to introduce Fourier series, orthogonal functions and applications of the Fourier method to boundary-value problems. 570 exercises. Answers and notes. 416pp. 5⅜ x 8½. 0-486-65973-9

COMPUTABILITY AND UNSOLVABILITY, Martin Davis. Classic graduate-level introduction to theory of computability, usually referred to as theory of recurrent functions. New preface and appendix. 288pp. 5⅜ x 8½. 0-486-61471-9

ASYMPTOTIC METHODS IN ANALYSIS, N. G. de Bruijn. An inexpensive, comprehensive guide to asymptotic methods–the pioneering work that teaches by explaining worked examples in detail. Index. 224pp. 5⅜ x 8½ 0-486-64221-6

APPLIED COMPLEX VARIABLES, John W. Dettman. Step-by-step coverage of fundamentals of analytic function theory–plus lucid exposition of five important applications: Potential Theory; Ordinary Differential Equations; Fourier Transforms; Laplace Transforms; Asymptotic Expansions. 66 figures. Exercises at chapter ends. 512pp. 5⅜ x 8½. 0-486-64670-X

INTRODUCTION TO LINEAR ALGEBRA AND DIFFERENTIAL EQUATIONS, John W. Dettman. Excellent text covers complex numbers, determinants, orthonormal bases, Laplace transforms, much more. Exercises with solutions. Undergraduate level. 416pp. 5⅜ x 8½. 0-486-65191-6

RIEMANN'S ZETA FUNCTION, H. M. Edwards. Superb, high-level study of landmark 1859 publication entitled "On the Number of Primes Less Than a Given Magnitude" traces developments in mathematical theory that it inspired. xiv+315pp. 5⅜ x 8½. 0-486-41740-9